Composites

This book emphasizes the importance of modeling in the initial design phase of a composite component. It covers a wide range of modeling techniques and multiphysics simulation using finite elements. It further provides practical examples and details studies that demonstrate the application of modeling techniques to real-world application of composite structures.

This book:

- Discusses manufacturing of different types of composite components using different techniques, and static and dynamic analyses of composites using FE modeling.
- Covers the machining performance of carbon nanotubes-reinforced nanocomposites and multiscale modeling techniques in composites.
- Presents a detailed study on ceramics matrix composite using modern machining operation, and hybrid nanocomposite using conventional machining operation.
- Highlights the development of hybrid nanocomposites and their tribological characteristics.
- Illustrates implementation of biomimicry for advanced impact resistance of composites and degradation of polyamides for future prospects.

It is primarily written for senior undergraduates, graduate students, and academic researchers in the fields of mechanical engineering, manufacturing engineering, materials science, production engineering, industrial engineering, and aerospace engineering.

Engineering Tribology, Manufacturing and Applied Energy

Series Editors
Yashvir Singh and Nishant Kumar Singh

About the Series
Tribology is the study of parts when they are in contact with each other and have a considerable impact during their sliding or rotating motion. The series will publish the books highlighting the latest developments and research in tribology with its application in the manufacturing and energy sector. The series will include books covering established areas of tribology and emerging fields such as tribo-chemistry, surface engineering, nano-tribology, and bio-tribology. The series will focus on the recent developments and advancements in green tribology for example environmentally friendly composites. Aimed at senior undergraduate, graduate students, academic researchers and professionals, the proposed series will cover the latest techniques in the field of tribology with applications in diverse fields such as manufacturing, energy, and medical.

Modeling and Simulation of Fluid Flow and Heat Transfer
Edited by Reshu Gupta, Mukesh Kumar Awasthi

For more information on series page, please visit our website: https://www.routledge.com/Engineering-Tribology-Manufacturing-and-Applied-Energy/book-series/CRCETMAE

Composites
Modeling and Manufacturing

Edited by
Vijay Kumar Singh
Nishant Kumar Singh
Yashvir Singh

CRC Press is an imprint of the
Taylor & Francis Group, an **informa** business

First edition published 2025
by CRC Press
2385 NW Executive Center Drive, Suite 320, Boca Raton FL 33431

and by CRC Press
4 Park Square, Milton Park, Abingdon, Oxon, OX14 4RN

CRC Press is an imprint of Taylor & Francis Group, LLC

ISBN: 978-1-032-74457-5 (hbk)
ISBN: 978-1-032-91646-0 (pbk)
ISBN: 978-1-003-56435-5 (ebk)

DOI: 10.1201/9781003564355

Typeset in Sabon
by KnowledgeWorks Global Ltd.

Contents

Contents

Preface

Composite materials have transformed many industries, including aerospace and automotive, by providing unique benefits such as high strength-to-weight ratio, corrosion resistance, and versatile manufacturing techniques. *Composites: Modeling and Manufacturing* delves into the complex realm of composites, covering both fundamental concepts and cutting-edge breakthroughs.

In Chapter 1, the basic concepts and kinds of composites are introduced. Metal matrix composites are discussed in Chapter 2 along with their evolution and applications. Novel production processes are covered in Chapter 3. In Chapter 4, the mechanical properties of individual nanocomposites are discussed, in relation to the metal matrix nanocomposites based on magnesium, carbon nanotubes, and aluminum oxide, as evaluated using the stir casting process. Chapter 5 delves into the ecological and financial advantages of using industrial waste into polymer composites. Chapter 6 focuses on computerized finite element modeling. In Chapter 7, we look at how sustainable composites are making use of natural fibers. The creation of coatings that are resistant to corrosion in marine environments is covered in Chapter 8. Chapter 9 delves into the topic of ceramic matrix composites, specifically looking at the advanced manufacturing processes and issues associated with producing these composites. In Chapter 10, the use of biomimicry in these types of composites is researched. Chapter 11 concludes with an examination of the processes and potential applications of polyamide degradation.

Researchers, engineers, and students interested in composite materials will find this book to be an excellent resource due to the breadth of topics covered, which range from basic ideas to sophisticated applications and sustainability. Editors are hopeful that this book will encourage more innovation and progress in this vital and ever-changing field.

Editor Biographies

Vijay K. Singh is an Assistant Professor, GR-I and Head of Mechanical Engineering Department at KK University, Nalanda, Bihar, India. He earned his Ph.D. from National Institute of Technology Rourkela (NITRKL) and B.Tech. in Mechanical Engineering from College of Engineering, Bhubaneswar, Odisha, India. He has published more than 27 research articles in well-known national/international journals/conferences and serves as Reviewer and Editorial Member of many peer-reviewed journals and conferences. His research interests include computational mechanics, Advanced composite structures, and smart materials and structures. He has also worked on several National level projects funded by DST, AICTE, MHRD, DRDO.

Nishant K. Singh is an Associate Professor at Harcourt Butler Technical University, Kanpur, Uttar Pradesh, India. He earned his Ph.D. from IIT, Dhanbad, and M.Tech. in Production Engineering and B.Tech. in Mechanical Engineering from Delhi College of Engineering. He has published more than 90 research articles in well-known international journals and serves as Reviewer and Editorial Member of peer reviewed journals and conferences. His research interests include tribology, composites, micro-manufacturing, and nonconventional machining processes.

Yashvir Singh is presently working as an Associate Professor in the Department of Mechanical Engineering, Harcourt Butler Technical University, Kanpur, Uttar Pradesh, India. He did his postdoctorate from Universiti Tun Hussein Onn Malaysia, Parit Raja, Malaysia. He earned his Ph.D. from the University of Petroleum and Energy Studies, Dehradun, Uttarakhand, India. He has published more than 110 research articles in various peer-reviewed journals. He is also reviewer and editorial board member of various journals. His specialization includes areas like tribology, biofuels, lubrication, and manufacturing.

Contributors

Alok Agrawal
Department of Mechanical
  Engineering
Sagar Institute of Research and
  Technology
Bhopal, Madhya Pradesh, India

Mohammad Asif Ali
Key Laboratory of Synthetic and
  Biological Colloids
Ministry of Education, School
  of Chemical and Material
  Engineering
Jiangnan University
Wuxi, China

P. M. G. Bashir Asdaque
Department of Mechanical
  Engineering, School of Engineering
Dayananda Sagar University
Bengaluru, Karnataka, India

Gagan Bansal
Department of Mechanical Engineering
IIT (BHU)
Varanasi, Uttar Pradesh, India
and
Department of Mechanical
  Engineering
Graphic Era (deemed to be)
  University
Dehradun, Uttarakhand, India

Sumit K. Bhowmik
Department of Mechanical
  Engineering
National Institute of Technology
  Silchar
Silchar, Assam, India

Nachiketa Das
School of Marine Engineering and
  Technology
Indian Maritime University, Kolkata
  Campus
Kolkata, West Bengal, India

Gaurav Gupta
School of Mechanical Engineering
Vellore Institute of Technology
Vellore, Tamil Nadu, India

E. Hemachandran
Department of Foundry and Forge
  Technology
National Institute of Advanced
  Manufacturing Technology
  (formerly NIFFT)
Ranchi, Jharkhand, India

Chetan Kumar Hirwani
Department of Mechanical
  Engineering
National Institute of Technology
  Patna, Bihar, India

Nitin Johri
Department of Mechanical Engineering
Graphic Era (Deemed to be) University
Dehradun, Uttarakhand, India

Bhaskar Chandra Kandpal
Department of Mechanical
  Engineering
Inderprastha Engineering College
Ghaziabad, Uttar Pradesh, India

Tatsuo Kaneko
Key Laboratory of Synthetic and
  Biological Colloids
Ministry of Education, School
  of Chemical and Material
  Engineering,
Jiangnan University
Wuxi, China

Ajay Kumar
Department of Mechanical
  Engineering
Inderprastha Engineering College
Ghaziabad, Uttar Pradesh, India

Rahul Kumar
Department of Mechanical
  Engineering, School of
  Engineering
Dayananda Sagar University
Bengaluru, Karnataka, India

Rajan Kumar
School of Marine Engineering and
  Technology
Indian Maritime University, Kolkata
  Campus
Kolkata, West Bengal, India

Rajesh Kumar
Department of Mechanical
  Engineering
National Institute of Technology
  Patna, Bihar, India

Ravi Kumar
Department of Mechanical
  Engineering
National Institute of Technology Patna
Patna, Bihar, India

Santosh Kumar
Department of Mechanical
  Engineering
IIT (BHU)
Varanasi, Uttar Pradesh, India

Natraj Mishra
School of Engineering
Adamas University
Kolkata, West Bengal, India

Sandip Kumar Nayak
Department of Mechanical
  Engineering
Indian Institute of Technology
New Delhi, India

Maiko Okajima
Key Laboratory of Synthetic and
  Biological Colloids
Ministry of Education, School
  of Chemical and Material
  Engineering
Jiangnan University
Wuxi, China

Akrom Palamanit
Department of Interdisciplinary
  Engineering
Faculty of Engineering
Prince of Songkla University
Hat Yai, Songkhla, Thailand

Pravat Ranjan Pati
Department of Mechanical
  Engineering
Graphic Era (Deemed to be
  University)
Dehradun, Uttarakhand, India

Abhilash Purohit
Department of Mechanical Engineering
Galgotias University
Greater Noida, Uttar Pradesh, India

Shreya Rai
Department of Aerospace Engineering
Indian Institute of Technology, Madras
Chennai, Tamil Nadu, India

Alok Satapathy
Department of Mechanical Engineering
National Institute of Technology
Rourkela, Odisha, India

Faladrum Sharma
Department of Mechanical Engineering
Indian Institute of Technology,
   Guwahati
Guwahati, Assam, India

Gaurav Sharma
Department of Physics
Meerut Institute of Engineering and
   Technology
Meerut, Uttar Pradesh, India

Sumit K. Sharma
Department of Metallurgical and
   Materials Engineering
BIT Sindri
Jharkhand, India

Amarish Kumar Shukla
School of Marine Engineering and
   Technology
Indian Maritime University, Kolkata
   Campus
Kolkata, West Bengal, India

Nand Kishore Singh
Department of Mechanical Engineering
Rowan University
New Jersey, USA

Nishant K. Singh
Department of Mechanical
   Engineering
Harcourt Butler Technical
   University
Kanpur, Uttar Pradesh, India

Pavitra Singh
Department of Foundry and Forge
   Technology
National Institute of Advanced
   Manufacturing Technology
   (formerly NIFFT)
Ranchi, Jharkhand, India

Vijay K. Singh
Department of Mechanical
   Engineering
K. K. University
Nalanda, Bihar, India

Yashvir Singh
Department of Mechanical
   Engineering
Harcourt Butler Technical
   University
Kanpur, Uttar Pradesh, India

Sandeep Tiwari
Department of Mechanical
   Engineering
National Institute of Technology
   Patna
Patna, Bihar, India

Sankata Tiwari
Department of Mechanical
   Engineering
IIT (BHU)
Varanasi, Uttar Pradesh, India

Chapter 1

A review on fundamental and structural properties of composite materials

Nishant K. Singh, Yashvir Singh, and Nand Kishore Singh

1.1 INTRODUCTION

The historical evolution of composite materials is a fascinating journey spanning decades, punctuated by advances in technology, materials science, and engineering. Composite materials are structures made up of two or more unique materials with differing physical or chemical properties that are combined to form a material having improved properties in comparison with parental materials. Composites have been around since antiquity; however, there has been substantial improvement in the recent age [1]. Composite materials are engineered materials composed of two or more constituent materials having distinctly different physical or chemical properties. When these materials are mixed, they form a new material with better and customized qualities compared to their components. Understanding composite materials' ingredients, production processes, characteristics, and applications is critical. Composite materials are important in many industries because of the unique combination of features acquired from distinct constituent materials [2]. Composite materials have a wide range of applications and significance, influencing industries such as aircraft, automotive, construction, and sports.

1.2 SIGNIFICANCE OF COMPOSITE MATERIALS

Composite materials have a higher strength-to-weight ratio than traditional materials such as metals. This makes them perfect for applications requiring weight reduction, such as the aerospace and automobile industries. Engineers can tailor the mechanical properties of composites by choosing and combining various materials [3]. This enables the customizing of strength, stiffness, and other characteristics to individual performance requirements. Composites are inherently corrosion-resistant, making them ideal for use in severe settings. Composites, unlike metals, do not rust or corrode, making them highly durable and long-lasting. Composite materials give designers more freedom in shaping and moulding components. This allows for the fabrication of intricate and aerodynamic structures that would not be possible with ordinary materials. Composites frequently display great fatigue resistance, making them ideal for applications involving repetitive loads, such as aerospace and automotive components [4]. Certain composite materials like carbon–carbon composite have exceptional thermal insulation qualities. This makes them useful in applications requiring temperature management, such as the construction of energy-efficient buildings. Composites can be developed with good electrical insulating qualities, making them valuable in electrical and electronic applications that require insulation.

1.3 APPLICATIONS OF COMPOSITE MATERIALS

Aircraft components, such as wings, fuselage sections, and interior structures, frequently use composite materials to reduce weight and increase fuel economy. Automotive manufacturers employ composites for a variety of components, including body panels, chassis parts, and interior elements, to reduce weight and increase fuel efficiency. Composite materials are used in the construction of bridges, buildings, and infrastructure due to their high strength, corrosion resistance, and design versatility. Composites are commonly used in sporting equipment such as tennis rackets, golf clubs, and bicycle frames due to their strength and weight benefits [5]. Composite materials are frequently employed in the production of wind turbine blades due to their strength, lightweight nature, and resistance to climatic conditions. Composites' corrosion resistance and great strength make them ideal for boat hulls and other marine components, ensuring endurance and lifespan in the water. Composite materials are employed in medical devices, prosthetics, and orthopaedic implants because they are biocompatible and have customized mechanical qualities. Composites with excellent electrical insulation qualities are used in the production of electronic components and insulation.

1.4 TYPES OF COMPOSITE MATERIALS

Composite materials are designed materials made up of two or more separate phases, each with its own set of properties, that are combined to produce improved performance or specialized qualities that individual components cannot achieve on their own. These materials are widely employed in a variety of industries due to their adaptability, strength-to-weight ratio, and customized features [6, 7]. There are various types of composite materials, each having its composition, properties, and applications. Here are a few common types.

1.4.1 Carbon fibre-reinforced composites

These composites are made up of carbon fibres embedded in a matrix substance, typically epoxy. Carbon fibre composites are well-known for their high strength, lightweight design, and good stiffness. They have uses in aeroplanes, automobiles, sporting equipment, and high-performance buildings.

1.4.2 Glass fibre-reinforced composites

Glass fibres are widely utilized to reinforce matrix materials such as polyester or epoxy resin. Glass fibre composites are low-cost, strong, and widely employed in construction, marine, automotive, and consumer products.

1.4.3 Aramid fibre-reinforced composites

Aramid fibres, including Kevlar, are well-known for their exceptional strength and impact resistance. Body armour, aircraft components, and sports equipment all employ aramid composites because they are tough and durable.

1.4.4 Metal matrix composites (MMCs)

These composites are made up of metal matrices (such as aluminium and titanium) that have been reinforced with ceramic particles or fibres. MMCs are desirable due to their excellent strength, thermal conductivity, and wear resistance. They have uses in aerospace, automobiles, and electronics.

1.4.5 Polymer matrix composites (PMCs)

These composites contain a polymer matrix reinforced with ceramic or metallic particles. PMCs are lightweight and resistant to corrosion, making them ideal for use in automobile components, sports equipment, and consumer goods.

1.4.6 Laminar composites

These composites consist of layers of reinforced fibres (often glass or carbon) embedded in a resin matrix. The stacking improves structural qualities and enables the optimization of strength and stiffness in certain directions. It is widely utilized in aviation components, automobile parts, and wind turbine blades.

1.4.7 Structural composites

These composites are made of a core material sandwiched between two outer layers. The core material is stiff and strong, while the outer layers give protection and additional structural support. These are commonly used in aerospace, marine, and automotive applications.

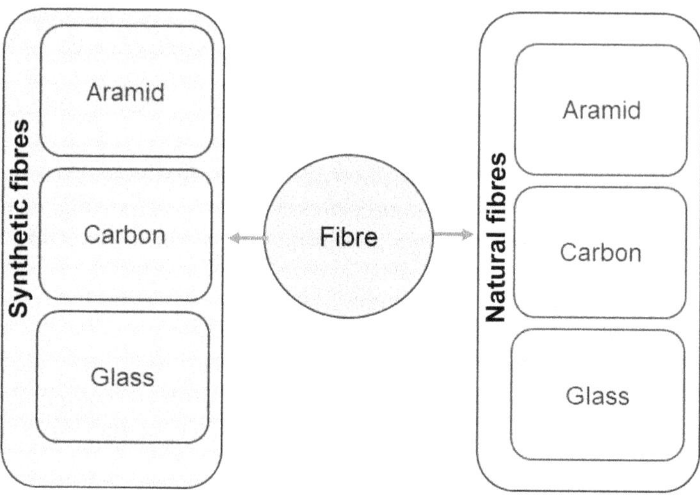

Figure 1.1 Commonly used fibres.

1.4.8 Ceramic matrix composites (CMCs)

Ceramic fibres or particles are embedded in a ceramic matrix. CMCs exhibit high-temperature resistance, making them suitable for applications in aerospace, gas turbine engines, and other high-temperature environments.

Composite materials are constantly evolving as researchers experiment with novel material combinations and manufacturing techniques. Engineers can adjust material qualities for individual applications, thanks to the vast spectrum of composite kinds, which contribute to technological and industrial breakthroughs. Composite materials are engineered materials composed of two or more constituent materials having distinctly different physical or chemical properties. The combination of these materials results in a new material with superior performance attributes to those of the separate components. The composition and structure of composite materials influence their qualities and applications. Figure 1.1 depicts the most common fibre used in composite manufacturing.

1.5 COMPOSITION OF COMPOSITE MATERIALS

1.5.1 Matrix material

The matrix is the major component of a composite material that surrounds and connects the reinforcement components. Depending on the desired qualities of the composite, it is commonly made of polymer, metal, ceramic, or a combination of these. Polymers such as epoxy, polyester, and phenolic resins are popular matrix materials because of their versatility, ease of processing,

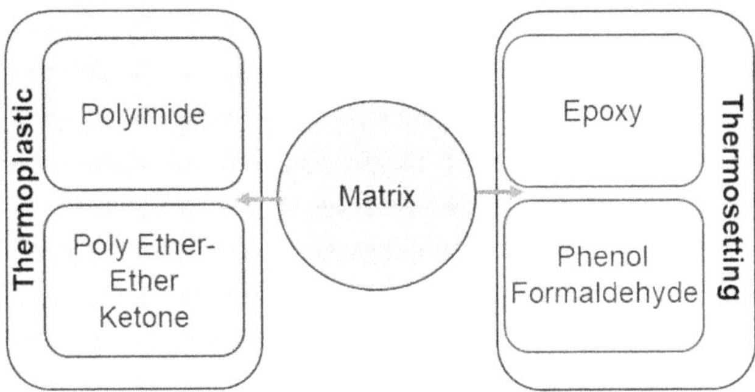

Figure 1.2 Commonly used polymeric matrix material for composite.

and compatibility with diverse reinforcements [8]. Figure 1.2 shows the most often used matrix material for composite production.

1.5.2 Reinforcement material

Reinforcement materials are incorporated into the matrix to improve certain qualities, including strength, stiffness, and toughness. Fibres (glass, carbon, aramid), particulate fillers (silica, alumina), and continuous strands are some of the most common reinforcement materials. The type of reinforcement material utilized depends on the application requirements, with carbon fibres being used for high-strength applications and glass fibres for cost-effective solutions [9].

1.5.3 Fillers and additives

Fillers like nanoparticles or microspheres may be added to improve specific properties, such as thermal conductivity, flame resistance, or wear resistance. Additives like plasticizers, stabilizers, or pigments can be incorporated to enhance processing, durability, or aesthetic qualities [10]. Table 1.1 shows the properties of composite-reinforcing fibres.

Table 1.1 Properties of composite-reinforcing fibres

Material	E(G)	σ_b (GPa)	ε_b (%)	P (Mg/m³)	E/ρ (MJ/kg)	σ_b/E (MJ/kg)	Cost (Rs./kg)
E-glass	72.4	2.4	2.6	2.54	28.5	0.95	61.6
S-glass	85.5	4.5	2	2.49	34.3	1.8	1,232–1,848
Aramid	124	3.6	2.3	1.45	86	2.5	1,232–1,848
Boron	400	3.5	1	2.45	163	1.43	18,480–24,640
HS graphite	253	4.5	1.1	1.80	140	2.5	3,696–6,160
HM graphite	520	2.4	0.6	1.85	281	1.3	12,320–36,960

Source: Ref. [11].

1.6 STRUCTURE OF COMPOSITE MATERIALS

The structure of composite materials influences the physical and mechanical properties of composites. The most widely utilized structure of composites is shown in Figure 1.3. In addition, various types of composite structures are discussed in detail in the following sections.

1.6.1 Particulate composites

Particulate composites contain tiny particles or fillers spread throughout the matrix material. The distribution and direction of these particles affect qualities such as hardness, wear resistance, and thermal conductivity.

1.6.2 Fibre-reinforced composites

Fibre-reinforced composites are made up of either continuous or discontinuous fibres embedded in a matrix. The fibre arrangement and orientation have a considerable impact on the composite's strength, stiffness, and other mechanical properties. Unidirectional, woven, and random fibre orientations are prevalent, with each providing significant benefits in particular applications.

1.6.3 Laminar composites

Laminar composites are made up of many layers (lamina) of different materials stacked together. The stacking sequence and orientation of these layers

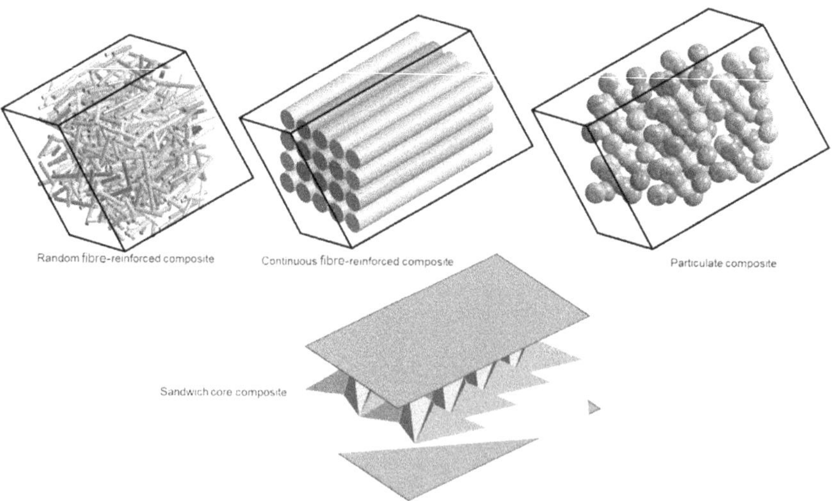

Figure 1.3 Commonly used structure of composite.

result in a composite with customized anisotropic characteristics, in which strength and stiffness vary with direction.

1.6.4 Structural hierarchies

Some sophisticated composite materials feature hierarchical structures that resemble natural materials such as bone or wood. These hierarchical structures help to improve strength, toughness, and damage resistance.

1.7 BENEFITS OF COMPOSITE MATERIALS

1.7.1 High strength-to-weight ratio

Composite materials frequently demonstrate great strength and stiffness while remaining lightweight, making them ideal for applications requiring significant weight savings.

1.7.2 Tailored properties

Engineers can customize composites for specific purposes by combining different materials and optimizing features like thermal conductivity, electrical conductivity, and corrosion resistance.

1.7.3 Durability and corrosion resistance

Composites are frequently resistant to corrosion and chemicals, which adds to their long-term resilience in tough situations.

1.7.4 Design flexibility

Composite materials provide design versatility, allowing for the production of complicated shapes and structures that would be difficult or impossible to achieve using standard materials. The composition and structure of composite materials are meticulously created to capitalize on the synergies between diverse materials, resulting in materials with superior qualities for a wide range of applications in industries such as aircraft, automotive, construction, and sports equipment. Ongoing research and improvements in composite technology are increasing the potential and adaptability of these materials [12].

1.8 MECHANICAL PROPERTIES OF COMPOSITE MATERIALS

Composite materials are engineered materials composed of two or more constituent materials having distinctly different physical or chemical properties. These materials are mixed to form a composite with qualities superior to the

Table 1.2 Mechanical properties of the main grades of glass fibre

Fibre	Density (kg/m³)	Young's modulus (GPa)	Virgin filament strength (MPa)	Roving strength (MPa)	Strain to failure (%)
Grade-A (alkali)	2,460	73	3,100	2,760	3.6
Grade-C (chemical)	2,460	74	3,100	2,350	–
Grade-D (dielectric)	2,140	55	2,500	–	–
Grade-E (electrical)	2,550	71	3,400	2,400	3.37
Grade-R (reinforcement)	2,550	86	4,400	3,100	5.2
Grade-S (strength)	2,500	85	4,580	3,910	4.6

Source: Ref. [13].

individual components. Composite materials' mechanical qualities are critical in determining their performance and application in a variety of sectors. Table 1.2 shows the mechanical properties of the main grades of glass fibre.

Here are some key aspects of the mechanical properties of composite materials.

1.8.1 Strength

Because of the mixture of many components, composite materials frequently have high strength. The arrangement and direction of the reinforcing fibres or particles within the matrix influence the overall strength of the composite. The tensile, compressive, and shear strengths of composite materials can be adjusted to fulfil specific technical needs.

1.8.2 Stiffness

Stiffness is a material's capacity to resist deformation under an applied stress. Composite materials can be quite stiff, making them ideal for applications that require rigidity and dimensional stability. The stiffness of composites is determined by the kind, orientation, and volume proportion of reinforcing components.

1.8.3 Flexural properties

Flexural strength and modulus are key properties for materials that are bent or flexed under load. Fibre matrix composites can have better flexural characteristics than separate constituents. The orientation and length of reinforcing fibres are critical in regulating the flexural performance of composite materials.

1.8.4 Fatigue resistance

Composite materials are intended to withstand fatigue, or the slow and cumulative damage caused by repeated loading and unloading cycles. Composite fatigue resistance is determined by matrix toughness and the ability of reinforcing components to efficiently distribute stresses.

1.8.5 Impact resistance

Impact resistance is critical for materials that may be subjected to unexpected forces. A composite material's ability to absorb and dissipate energy upon impact is determined by the qualities of both the matrix and the reinforcing elements. The toughness of the matrix material is critical in defining a composite's impact resistance.

1.8.6 Creep resistance

Creep is the slow distortion of a material due to a steady load over time. Composite materials can be manufactured to have low creep, making them ideal for applications requiring dimensional stability.

1.8.7 Density

Composite materials often have a lower density than traditional materials such as metals. This produces lightweight structures, making composites appealing for applications requiring weight reduction without sacrificing strength and stiffness.

1.8.8 Thermal properties

The thermal expansion and conductivity of composite materials are determined by the characteristics of the matrix and reinforcing elements. Engineering composites with precise thermal characteristics is critical for applications where temperature fluctuations are a problem.

Understanding and adapting these mechanical properties enables engineers to create composite materials for a variety of applications, including aircraft, automotive, construction, and sports equipment. Composite materials are versatile because they may be customized to meet the needs of a specific application.

1.9 MANUFACTURING PROCESSES

Composite materials are manufactured by combining two or more different components to form a new material with superior qualities. Composites are commonly utilized in a variety of industries, including aerospace, automotive, construction, and sports equipment, due to their high strength-to-weight ratio, corrosion resistance, and adaptability. Composite materials are typically manufactured using the following processes.

1.9.1 Material selection

The matrix material binds the reinforcement together and distributes loads among the reinforcing fibres. Common matrix materials include polymers (thermosetting or thermoplastic), metals, and ceramics. Reinforcements, typically fibres or particles, improve the mechanical properties of the composite. Glass fibres, carbon fibres, aramid fibres, and natural fibres such as jute or bamboo are among the most common reinforcement materials.

1.9.2 Layup

Layup is the process of aligning layers of reinforcing and matrix materials in a specified orientation to obtain the appropriate mechanical characteristics. Automated machinery or expert workers manually layer the materials in a specified arrangement.

1.9.3 Impregnation

The matrix material is applied to the reinforcing layers in a liquid or semi-liquid state to fully wet and wrap the fibres. This might be accomplished using procedures such as hand layup, spray layup, or vacuum-assisted resin transfer moulding (VARTM).

1.9.4 Curing (polymerization)

The curing process in thermosetting composites involves the polymerization of the resin, which results in a stiff and stable structure. This can be accomplished via heat, chemical processes, or a mix of the two, depending on the type of resin utilized.

1.9.5 Moulding

The layup is placed in a mould, which determines the final shape of the composite item. Common moulding processes used in composite manufacturing include compression moulding, injection moulding, and resin transfer moulding (RTM).

1.9.6 Consolidation

After moulding, the composite goes through a consolidation procedure to ensure that the fibres and matrix are well connected. This may include using heat and pressure to further cure the resin and eliminate any voids in the material.

1.9.7 Finishing

To meet specific dimensions and aesthetic requirements, the completed composite product may be subjected to further processing such as trimming, machining, or surface treatment.

1.9.8 Quality control

Quality control methods are applied throughout the manufacturing process to guarantee that the finished product satisfies the required requirements. Non-destructive testing methods, visual inspections, and other techniques are used to detect flaws or anomalies.

1.9.9 Post-curing (if necessary)

Some composite materials may require additional post-curing to enhance their properties further.

1.9.10 Assembly (if necessary)

Composite parts can be combined with other components to form larger constructions or final products.

These production procedures increase the versatility and efficiency of composite materials, making them more popular in an array of fields across sectors.

1.10 CHARACTERIZATION AND TESTING OF COMPOSITE MATERIALS

Characterization and testing of composite materials are critical for ensuring their performance and dependability in a variety of applications. Composite materials, which are made up of two or more distinct phases with varying properties, provide several desired characteristics, including a high strength-to-weight ratio, corrosion resistance, and customized mechanical properties [14]. Proper characterization and testing methodologies are essential for understanding and optimizing the behaviour of these materials.

1.10.1 Characterization

- The initial stage in composition analysis is to identify elements and their proportions. This includes techniques such as spectroscopy, X-ray diffraction, and elemental analysis.
- Microstructural analysis: Microscopy techniques such as optical microscopy, scanning electron microscopy (SEM), and transmission electron microscopy (TEM) can reveal internal structure, fibre/matrix distribution, and probable flaws.
- Mechanical properties: Determining mechanical properties such as tensile strength, compressive strength, shear strength, and modulus of elasticity offers information on the material's structural integrity. This includes standardized testing methods such as ASTM or ISO protocols.
- Thermal analysis: Understanding a material's sensitivity to temperature fluctuations is critical. Differential scanning calorimetry (DSC) and

thermogravimetric analysis (TGA) aid in determining thermal stability, glass transition temperature, and degradation points.

- Composites must be tested for chemical resistance in harsh settings. Immersion testing and exposure to specific chemicals aid in determining resistance.
- Composite materials may have electrical applications. Characterizing their electrical conductivity or resistivity is crucial. Four-point probe measurements are among the techniques used.
- Non-destructive testing (NDT) involves using procedures such as ultrasonic testing, radiography, or thermography to assess the interior integrity of composite structures without causing damage.

1.10.2 Testing

- Tensile testing analyses a material's strength and deformation under axial loading.
- Compression testing assesses a material's capacity to endure compressive loads, which is important for applications with high compression forces.
- Flexural testing evaluates a material's resistance to bending, revealing structural performance under various stress circumstances.
- Shear testing evaluates a material's capacity to sustain forces parallel to its surface, which is critical for shear stress applications.
- Impact testing measures a material's toughness under rapid loading conditions. Charpy and Izod tests are commonly used to determine impact resistance.
- Fatigue testing evaluates material endurance under cyclic loading, imitating real-world stress scenarios.
- Environmental testing involves exposing the composite to various environmental conditions, such as temperature, humidity, and UV exposure, to determine its long-term stability.
- Fracture toughness testing evaluates a material's capacity to resist fracture propagation, ensuring reliability and safety.

To guarantee a thorough understanding of composite materials' properties and behaviour, they must be characterized and tested using a multidisciplinary approach. This information is critical for designing and producing high-performance composite structures in a variety of industries, including aerospace, automotive, and construction.

1.11 DESIGN CONSIDERATIONS

When designing using composite materials, a variety of elements must be considered to optimize performance, durability, and efficiency. Composite materials are made up of two or more separate components that, when mixed, produce a material with improved qualities. Here are the main design considerations for composite materials:

- Materials selection: The matrix, or basic substance, holds the reinforcements together. Polymers, metals, and ceramics are among the most commonly used matrix materials. The choice is based on the application, taking into account aspects like strength, weight, and cost. Reinforcements, such as fibres or particles, add strength and stiffness to the composite. Fibres such as carbon, glass, and aramid are commonly used. The direction and type of reinforcement have an impact on the material's mechanical characteristics.
- Performance requirements: Consider the necessary strength, stiffness, and toughness for the application. Customize the composite's composition and structure to satisfy these mechanical specifications. Evaluate the material's behaviour at various temperature conditions. Some composites may be better suited to high-temperature applications, while others thrive in low-temperature conditions.
- Composite materials offer excellent strength while being lightweight. Design considerations should include optimizing the density of the material to obtain the necessary strength-to-weight ratio.
- Understand the composite material's fatigue resistance under cyclic loading circumstances. Design for fatigue resistance to ensure longevity, particularly in applications with recurrent stress or dynamic loads.
- Environmental considerations: Determine the composite's resistance to environmental variables such as moisture, chemicals, UV radiation, and corrosion. Some composites may require additional coatings or treatments to improve durability in certain situations.
- Manufacturability: Evaluate the composite material's ease of manufacture and processing. The chosen design should be compatible with the chosen manufacturing methods, such as moulding, filament winding, pultrusion, or other techniques.
- Cost considerations: Determine the total cost-effectiveness of the composite material for the proposed application. While composites can provide higher performance, it is critical to weigh their advantages against the costs of materials, production, and any additional processing.
- Understand how the composite material will be connected or bonded with other components. This is critical to the structural integrity of the finished product. Adhesive bonding and mechanical fastening are two techniques that could be used.
- Implement quality control procedures during manufacture to ensure consistent and reliable composite components. This includes tracking material parameters, production processes, and final product inspections.
- Evaluate the product's whole life cycle, from production to disposal. Consider recycling options and environmental impact, following sustainable design standards.

Incorporating these design concerns enables engineers and designers to produce composite materials that meet particular performance objectives while addressing practical limits and assuring cost-effective manufacturing.

1.12 APPLICATIONS

Composite materials offer strength, lightweight, durability, and versatility, making them widely used in a variety of sectors. Some significant applications of composite materials are as follows:

- Composite materials, such as carbon fibre-reinforced polymers (CFRP), are widely utilized in aircraft manufacturing for wing, fuselage, and tail surfaces. These materials have a high strength-to-weight ratio, which improves fuel efficiency and overall performance.
- Composite materials are commonly utilized in automotive manufacturing for structural components like body panels, chassis, and interiors. Their lightweight design reduces vehicle weight, improving fuel efficiency and overall performance.
- Carbon–ceramic composites are used in high-performance brake systems due to their superior heat resistance and lightweight.
- Composite materials are commonly utilized in the marine sector to create boat hulls, decks, and structural components. They provide excellent strength and corrosion resistance, which extends the life of marine constructions.
- Composite materials are used to manufacture bridge components such as decking and girders. They offer great strength, corrosion resistance, and low maintenance requirements.
- Fibre-reinforced composites are used in concrete to improve strength and durability, especially in applications where weight is an important consideration.
- Composite materials, particularly carbon fibre, are commonly utilized to manufacture sports equipment like bicycles, tennis rackets, golf clubs, and hockey sticks. The materials' lightweight and high-strength properties help to increase performance.
- Composite materials are used in wind turbine blades for their strength and flexibility. The lightweight characteristic of composites allows for larger and more efficient blades, which contributes to higher energy output.
- Composite materials are commonly utilized in consumer electronics due to their lightweight and durable qualities, including smartphones and laptops. Carbon fibre, in particular, is employed in laptop enclosures and smartphone components.
- Composite materials are utilized in both professional and recreational sporting equipment, including skis, snowboards, and fishing rods.

- Composite materials are used to create lightweight and durable prosthetic limbs and orthotic devices that enhance user comfort and functionality.
- Composite materials are utilized to create lightweight and durable armour systems for military vehicles and people protection.
- The oil and gas industry uses composite materials for pipelines and storage tanks due to their resistance to corrosion and harsh environmental conditions.
- The ongoing development of novel composite materials and production technologies is set to broaden their uses even further, making them an essential component of many industries requiring high-performance materials with specialized qualities.

1.13 CHALLENGES AND FUTURE TRENDS OF COMPOSITE MATERIALS

Composite materials, which have been engineered to incorporate two or more materials with substantially distinct physical or chemical properties, have gained popularity in a variety of industries due to their distinctive characteristics such as high strength-to-weight ratio, corrosion resistance, and design flexibility. However, they confront numerous obstacles and are prone to changing trends as technology improves. Here, we look at the issues and future developments linked with composite materials.

1.13.1 Challenges

- High-performance fibres and resins for composite materials can be costly.
- Advanced manufacturing procedures, including autoclave curing, might lead to higher production costs.
- Environmental impact: Composite materials are difficult to recycle, raising environmental issues.
- Waste disposal: Traditional recycling procedures are sometimes ineffective for end-of-life disposal, leading to common difficulties.
- A lack of standardized testing and certification procedures can limit uptake.
- Quality assurance: Maintaining uniform quality among composite materials is challenging.
- Understanding and predicting the long-term behaviour of composite materials, particularly in severe settings, is a challenge.
- Designing composites to survive cyclic stress and fatigue is a continuous concern.
- Developing effective bonding and adhesion technologies for composite parts without compromising structural integrity is challenging.
- Successful integration requires compatibility with established materials and production methods.

1.13.2 Future trends

- 3D printing: Increasingly used for composite structures, additive manufacturing enables elaborate designs and reduces waste.
- Implemented automation and robots in production to improve efficiency and lower labour costs.
- Self-healing materials: Use self-healing technology to improve durability and save maintenance costs.
- Functionalized materials: Integrating sensors and actuators into composite structures for real-time monitoring and adaptive functionality.
- Bio-based materials: Creating composites from renewable and bio-based sources to reduce environmental impact.
- Improved recycling methods: New recycling methods make composite materials more environment-friendly.
- Nanotechnology integration: Using nanoparticles to enhance mechanical, thermal, and electrical qualities.
- Nanocomposites provide lightweight, strong, and multifunctional structures, leading to improved performance.
- Utilizing digital twin technology for virtual testing and predictive modeling to optimize designs.
- Data-driven design: Using big data and AI to improve material design and production processes.
- Industry standards: Developing standardized testing and certification methods.
- Developing regulatory frameworks for composite materials in crucial applications.

While composite materials have enormous promise for innovation in a variety of industries, tackling obstacles such as cost, recyclability, and durability is critical. Future composite material developments include improving production procedures, including smart features, increasing sustainability, investigating nanocomposites, and embracing digital technology to move the sector ahead. The chapter covers the principles of composite materials, including their types, composition, mechanical properties, manufacturing techniques, characterization, environmental effects, design considerations, applications, and potential future trends. Each part can be supplemented with appropriate facts, examples, and visuals to help readers understand the material more deeply.

REFERENCES

1. Gay D. Matériaux Composites. Editions Hermes, Paris, 1991
2. Schwartz MM. Composite Materials Handbook. McGraw Hill Book Co., 1983.
3. Strong AB. Fundamentals of Composites Manufacturing: Materials, Methods and Applications. Society of Manufacturing Engineers, Dearborn, 1989
4. Todor MP, Bulei C, Kiss I. Systematic approach on materials selection in the automotive industry for making vehicles lighter, safer and more fuel-efficient. Appl Eng Lett 2016;1(4): 91–97

5. Todor MP, Bulei C, Kiss I. An overview on fibre-reinforced composites used in the automotive industry. ANN Fac Eng Hunedoara: Int J Eng 2017;XV(2) 181–188

6. Yuanjian T, Isaac DH. Impact and fatigue behaviour of hemp fibre composites. Compos Sci Technol 2007;67:3300–3307. DOI: 10.1016/j.compscitech.2007.03.039

7. Williams J. The Science and Technology of Composite Materials. Australian Academy of Science, 2019. Available from: https://www.science.org.au/curious/technology-future/composite-materials

8. Brent Strong A. Fundamentals of Composites Manufacturing: Materials, Methods, and Applications. 2nd ed. Society of Manufacturing Engineers, 2008.

9. Saheb DN, Jog JP. Natural fibre polymer composites: A review. Adv Polym Technol 1991;18:351–363.

10. Jawaid M, Abdul Khalil HPS. Cellulosic/synthetic fibre reinforced polymer hybrid composites: A review. Carbohydr Polym 2011;86(1):1–18. DOI: 10.1016/j.carbpol.2011.04.043

11. Gerstle FP. Composites. Encyclopedia of Polymer Science and Engineering. New York: Wiley, 2009

12. Tarnopol'skii YM, Peters ST, Beil AI. Filament winding. In: Peters ST, editor. Handbook of Composites. Boston, MA: Springer, 1998.

13. Available from: http://d2n4wb9orp1vta.cloudfront.net/resources/images/cdn/cms/0409ct_Glassfiber_3.jpg.

14. Mohanty AK, Misra M, Hinrichsen G. Biofibres, biodegradable polymers and biocomposites: An overview. Macromol Mater Eng 2000;276–277(1):1–24.

Chapter 2

Progress and processing routes of metal matrix composites with their applications: A review

Yashvir Singh, Nishant Kumar Singh, and Akrom Palamanit

2.1 INTRODUCTION

The distribution and interface between the matrix and reinforcement, in addition to the size, volume percentage, shape, and composition of reinforcements, affect the performance of metal matrix composites (MMCs) [1]. Traditional methods for producing composites include additive manufacturing, casting, and spray forming. However, studies conducted over the past ten years have shown that powder metallurgy (PM) techniques account for about 28% of large-scale industrial production of MMCs [2].

In MMCs, design configuration refers to the interfaces, phases, and microstructural elements that are arranged, organized, and geometrically laid out within the composite material. With this method, particular composite structures that support required mechanical and functional qualities are to be created. Grain size, interface bonding, overall shape, and the distribution of reinforcing phases are some of the components that make up the design configuration. The composite can be made to match the specifications of its intended application by customizing design configurations to maximize the ratio of strength to toughness or other desired attributes [3, 4].

Al and Mg are the most frequent metals that are available for MMC applications. Significant attention has been drawn to magnesium-based composites due to their superior mechanical qualities compared to monolithic alloys. Nevertheless, a few drawbacks have slowed the advancement of magnesium use in autos. The low resistance to fracture and low ductility are the main causes. At high temperatures, magnesium is very reactive. Surface coatings or its naturally occurring oxide, however, can control it. To prevent oxidation with the environment, an inert atmosphere should be maintained during the manufacturing of magnesium-based MMCs [5, 6]. When iron is used as the matrix, it has a lot less impact strength and is more brittle than composite materials. Consequently, steel-based MMCs only exhibit significant promise in applications requiring resistance to wear. Applications in maritime environments are not appropriate for it. The primary applications for copper-based MMCs are those in which the properties of electrical and thermal

DOI: 10.1201/9781003564355-2

conductivity are important. Because of its low strength, pure Cu is not suitable as a matrix for many applications. Aluminium and its alloys are among the several matrix materials that are readily available and are frequently utilized to create MMCs. Aluminium has several desirable qualities, including being lightweight, economically viable, simple to work using many methods, and having incomparable corrosion opposition [7, 8].

Particulates, fibres, or even deep penetrating types could be applied as reinforcements. Composites can be categorized as fibre-reinforced composites, laminar composites, flake composites, filled composites, and element-reinforced composites based on the category of reinforcement utilized. Particulate reinforced composites are the main topic of this chapter as they are more widely available, minimum in cost, easier to distribute evenly throughout the matrix, and more easily dispersed. Established on the goals and intended uses of the composite, material for reinforcement were chosen. Graceful metal reinforcing allows for applications where weight reduction is of utmost importance. One of the most often used MMCs that results in improved mechanical qualities at comparatively cheaper production costs is Al reinforced with SiC [9], Al_2O_3 [10], or B_4C [11]. As a result, a lot of engineers are drawn to using aluminium metal matrix composites (AMMC) for a variety of applications, including drive shafts, cylinder liners, brake rotors, and pistons. When fabricating composite materials, bonding in the materials at the interface is a major source of concern. The produced composites may not have the desired qualities if the matrix and reinforcing components are not properly matched [12].

With its low weight, high strength, and high toughness, lightweight advanced metal provides a tactical improvement, especially in applications involving electronics, electric vehicles (EVs), and aircraft. Advanced aluminium matrix composites (AMCs) are acknowledged as fundamental building blocks for modern spacecraft development and aerospace production, allowing for the development of multifunctional, cost-effective solutions that simultaneously meet demanding standards for sustainability and fuel efficiency [13, 14]. This cutting-edge method of aircraft engineering offers trustworthy technical solutions to deal with present problems and satisfy upcoming design specifications. Thermal management, ground transportation (automotive and rail), aerospace, industrial, recreational, and infrastructure sectors are just a few of the industries that have benefited from the noteworthy belongings of metal composites, which also include great effectiveness, outstanding wear opposition, and appropriate electrical and thermal features. The structure's weight is reduced by around 20–25% and its energy strength is increased by about 15–20% when high strong point and durability aluminium composites are used in place of traditional aluminium alloy materials [15].

Aerospace applications that operate in harsh environmental circumstances, such as low temperatures, alternating temperatures, and high overloads, have a lot of potential for using AMCs. AMCs present an intriguing alternative in sectors such as aircraft and EVs, where reducing weight is essential to reducing fuel use and carbon emissions. Manufacturers are able to tackle the task of dropping weight and fuel usage while preserving dependability and cutting costs while retaining structural integrity. When AMCs are used in conjunction with axial flux electric motors, it is feasible to reduce the rotor weight by an astounding 40%, which increases the rotor's potential power. Furthermore, AAMCs have exceptional heat resistance and can tolerate temperatures as high as 300°C. AMCs are a better material than other composites for a variety of applications, including batteries, fans, and flywheels, because of their superior thermal resistance [7, 12].

The maximum complex and hard work in the design, manufacture, and treating of composites is, without a doubt, the design configuration. As long as exact control over the structural parts is attained, this concept's design arrangement offers a viable means of concurrently enhancing the power and durability of materials from a microstructural standpoint. In order to further investigate this problem, it is imperative to recognize that localized stress focus and distortion disparity are the root causes of the strength–toughness inversion connection in MMCs [16]. Because stress attention is more likely to occur at the grain edge, the high density also becomes the preferred location for micro stage throughout the failure process. The predominant approach to material design in conventional metal composites has been to achieve "homogenization and recombination" of a single-phase reinforcement. In order to do this, a number of problems are addressed, including insufficient matrix density, nano-reinforcement agglomeration, surface reactions, and various flaws. Improving the strength–toughness inversion relationship in MMCs is the goal. Biological composites intention in manmade composites has undergone significant development and evolution over time, leading to the emergence of distinctive composite configurations. These natural composites have amazing qualities; one prominent example of a biological structural composite that can support weight is bone. Bone has remarkable strength, durability, lightweight, and remarkable self-healing power. The natural evolution of micro–nano bricks through structural recombination is responsible for the notable increase in strength and fracture energy. Bone's composite nature made up of both soft and strong structure hunks and its graded organization across several scale length are the main causes of this remarkable performance. On the other hand, the design configuration of synthetic materials presents difficulties because little is known about the ways in which structural variables combine to improve strength and toughness, and there are insufficiently efficient processing techniques to get accurate control over the multilevel structures [17, 18].

In order to achieve this goal, this chapter attempts to present a thorough analysis of this design, giving a synopsis of its unique mechanisms for toughening and strengthening. This chapter presents different design processes for MMCs and discusses the challenges with the processes applied.

2.2 METHODS FOR THE DEVELOPMENT OF MMCs

The following are the processing routes for development of MMCs:

- Solid-state processing
- Liquid-state processing
- In situ processing

2.2.1 Solid-state processing

The technique of creating MMCs in a solid state at high temperatures and pressures involves joining the matrix metal and dispersion phase through mutual diffusion. This is known as solid-state fabrication.

2.2.1.1 High-energy ball milling

Powders are treated by high-energy ball milling, which leverages the milling media's action to mix, scatter, activate, and create composite structures. There are numerous designs for the milling devices. They are often separated into two categories: mills with high and low energy. Deformation, defect accumulation, fracture, welding, structural refinement, and the breaking and/or formation of chemical bonds are the processes that take place during high-energy ball milling. These are the particles that have every ingredient from the initial powder mixture. Due to a partial contact, they might also contain the reaction products; the reaction yield varies based on the milling energy and time. The consolidation behaviour and reactivity of composites are dictated by their internal structure. When consolidating materials properly, the bulk state of the material's non-equilibrium structure can be maintained. Numerous reviews on mechanical milling have been published, with varying emphasis on the technology or its uses [19, 20]. The Suryanarayana and Nasser Al-Aqeeli [21] provided a comprehensive analysis of nanocomposites made using mechanical milling. One benefit of mechanical milling for processing composite materials, they thought, was the ability to add a large volume proportion of reinforcing particles to the matrix. Hadef [22] evaluated the synthesis of intermetallics by high-energy ball milling. An overview of the usage of high-energy ball milling for the creation of powders with varying chemical compositions and natures, such as amorphous and nanocrystallite powders, was given by Zhang [23]. The main elements influencing the powders' sintering behaviour were thought to be the irregular form of the particles and the high degree of strain hardening of the high-energy ball milling products. Figure 2.1 shows the mechanism of milling.

2.2.1.2 Microwave sintering

Due to the shallow penetration depths, metals are known to reflect microwaves and produce sparks. This has given rise to a myth that metals cannot be treated with microwaves, which is somewhat accurate in the case of sintered or

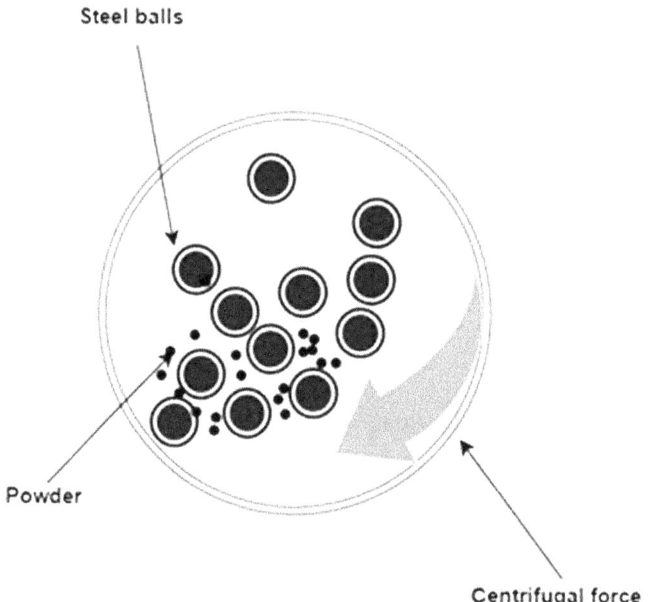

Figure 2.1 Process of milling using high energy balls.

monolithic metals. But when the metal is in the powder form, microwave radiation can be disseminated, and can reach deep into a sample helping to heat and sinter it. This property may be useful for additive technologies [24]. Roy [25] conducted a groundbreaking investigation on the sintering of pure powders of iron, cobalt, and alloys based on these metals. After adding a few percent of organic binder, the powders were cold-pressed to create "raw" billets with rectangular and cylindrical shapes. A custom-built furnace was used to sinter samples of different sizes and shapes, allowing the temperature to be adjusted from room temperature to 2,000°C. Thermal insulation was utilized in order to stop heat losses. The insulator did not absorb microwave radiation at low temperatures [26]. Several atmospheres were used for sintering, including hydrogen, air, and forming gas. They employed "susceptors" (SiC and $MoSi_2$ rods), which are highly effective in absorbing electromagnetic energy and converting it to heat, to enhance the heating conditions. Perhaps the only difference between the process with and without susceptor utilization is the ability to significantly raise the process rate [27]. The method of sintering bronze powder in a protected environment was investigated by Sethi et al. [28] using combined convection and microwave heating. Consequently, it was determined that, in comparison to convection sintering, microwave heating causes the bronze powder to sinter considerably more quickly. When compared to samples acquired through convection sintering, the bronze samples obtained using microwave sintering have a higher degree of hardness. The technique of microwave sintering a mixture of powder and pure copper powder was examined by the Takayama et al. [29].

According to Takayama et al. [29], the density of the raw billet determines when the volumetric heating of the materials begins, which also happens to be the beginning temperature of sintering. They examined the microhardness values during microstructural investigations and concluded that material compaction begins at the surface. The gradient in microhardness values can be eliminated by lengthening the soaking time at the sintering temperature. It was also discovered that there was no anticipated reduction in the sample's ability to permit microwave radiation to pass through, even in the presence of the creation of a metal contact between powder particles in the low-porosity raw billet. This process begins when the powder particles come into touch with one another at high temperatures, known as necking. Consequently, it was determined that materials with mechanical and physical characteristics similar to those sintered by convection technique could be obtained by microwave sintering [30]. The microwave sintering copper powder in argon, nitrogen, and forming gas (mixture of $3H_2 + N_2$) were among the protective atmospheres used in Mahmoud et al. [31]. Implementing a forming gas results in best sintering. Figure 2.2 shows the schematic image of the microwave sintering process.

2.2.2 Liquid-state processing

The process entails incorporating a dispersed phase into a molten matrix metal and then solidifying it. A high degree of mechanical characteristics in the composite will depend on the dispersed phase and liquid matrix having good interfacial bonding, or wetting. Coating the particles in the dispersed phase can improve wetting. In addition to lowering interfacial energy, appropriate

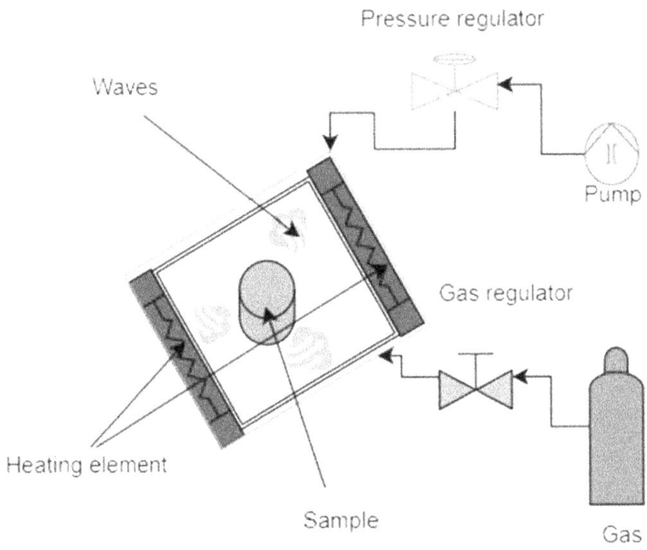

Figure 2.2 Microwave heating mechanism during sintering.

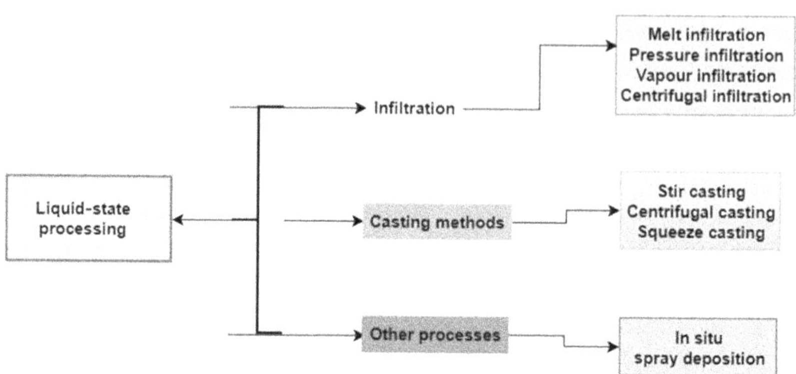

Figure 2.3 Liquid-state processing methods.

coating stops the dispersed phase and matrix from reacting chemically [32]. Figure 2.3 shows the processes involved in liquid-state processing.

2.2.2.1 Infiltration

The process of infiltration involves the penetration of molten metal into a preform by means of meltor pressure infiltration. In the case of melt infiltration, material for reinforcement are primary inserted into the die, after which the molten alloy is pierced onto it and permitted to freeze deprived of the application of outside pressure. In the case of pressure infiltration, pressure is put either straight or done by a combination [33, 34].

2.2.2.1.1 Molten infiltration

Figure 2.4 shows the molten infiltration processing route of MMCs. The most popular method of processing MMCs among the primary applied

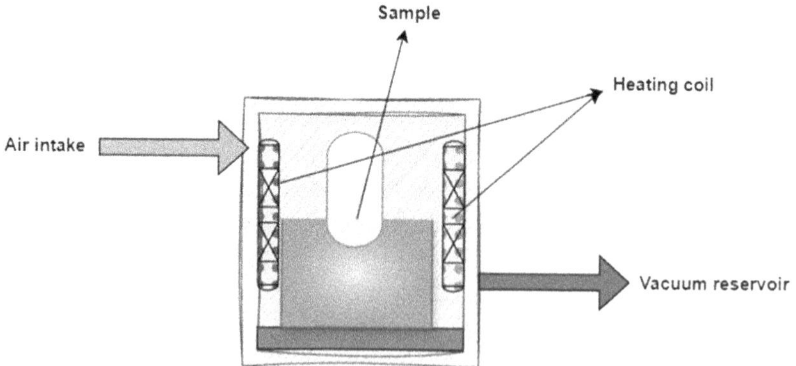

Figure 2.4 Melt infiltration processing route of MMCs.

manufacturing techniques is known as the "molten infiltration technique". Generally speaking, there are three basic steps to infiltration methods: firstly, a mould is filled with a bed of particles; secondly, a metal block is melted and liquid metal is injected into the porous bed; and thirdly, the sample is allowed to cool to room temperature either naturally or artificially. Several investigations have found that the metal alloys used in this approach typically have low melting points. Despite this, recent advances in liquid metal infiltration technology have made it possible to use this technique to produce syntactic foams from metal alloys with low to medium melting points (like aluminium, zinc, magnesium, etc.) as well as high melting points (like glass, iron, titanium, etc.) [35]

Compared to PM procedures, the filling process of molten infiltration technologies poses a significant disadvantage. The filler content ranges from 50 vol% for the random close packing sphere to more than 70 vol% for the densest regular packing, which follows a face-centred cubic cell, also known as cubic close-packed (CCP), for spherical particles of equal size. As a result, processing syntactic foams with a low volume percentage of fillers (<50 vol%) has significant challenges. The interstitial space of the preform determines how much matrix material is needed, hence this method is typically not used to synthesize syntactic foams with high matrix volume fractions. Furthermore, metal matrix syntactic foams processed by molten infiltration have strict shape constraints since infiltration moulds cannot accommodate complex geometries. Nonetheless, metal matrix syntactic foams with the maximum volume fraction of fillers can be processed. When compared to the other processing methods, this is by far one of the most important benefits [36].

2.2.2.1.2 Pressure infiltration

A prefabricated dispersion phase is soaked in a molten matrix metal in pressure infiltration process methods, which fills the gaps between the dispersed reinforcing phases. The capillary force of the dispersed phase or an external pressure applied to the liquid matrix phase (forced or pressure infiltration) can be the driving factor behind the infiltration process. The process of pressure infiltration has become significant from a business standpoint. Aluminium is the most often utilized lightweight structural material and is also the most widely used matrix material. For almost 30 years, pressure infiltration has been used to mass produce reinforced Al-Al_2O_3 automotive engine components, including diesel engine pistons, engine block cylinder liners, and crankshaft pulleys [37]. Another well-known use of the method is the production of tungsten-reinforced copper for electrical connections. The pressure infiltration approach, according to Guo et al. [38], exhibits superior thermal conductivity because the interface connection between diamond and aluminium with 12% Si composites is strengthened. Aluminium with 12% Si and graphene composites were manufactured by Narciso et al. [39] and showed good mechanical and thermal properties, making them appropriate for use in piston engine production.

2.2.2.1.3 Vapor infiltration

A thin coating that serves as the interphase of whisker-reinforced composite can be formed on the surface of the whiskers to cause fracture deflection or stop an interfacial reaction. A versatile technique for depositing interphase at moderate temperatures (600–1,200°C) is vapor infiltration. The borazine interphase is amorphous because the deposition temperature is often lower than the crystallization temperature of these materials. The ideal interphase materials are hypothesized to have a layered crystal structure. In order to convert the amorphous borazine interphase into a layered crystal structure, a very high-temperature heat treatment is required generally above 1,550°C [40]. If the transition from amorphous phase to crystal or interphase is carried out, the ceramic fibre in ceramic fibre-reinforced composite causes major degradation. Because of the exceptional heat stability of whiskers, the operation is permitted for composites reinforced with whiskers. Nonetheless, a unique instance surfaced lately when Zou et al. [41] devised an innovative vapor infiltration approach to generate well-oriented h-BN interphase at 1,200°C. In this instance, a bubbling technique was used to introduce borazine, a precursor of borazine, vapour into the reaction chamber. The formation mechanism of h-BN at such low temperatures remains unknown. For whisker preforms with extremely small pore diameters, the concurrent vapour infiltration and reaction during the process may produce an inhomogeneous interphase, despite the fact that vapour infiltration is an extremely flexible method. This has the potential to compromise the mechanical integrity of the composites. In addition, further investigation is warranted to ascertain the microstructure, composition, and thickness of the interphase within a specific composite reinforced with whiskers. Furthermore, research has shown that interfacial weak bonding in fibre-reinforced composites results in an exceptionally high fracture toughness. However, the interfacial weak or strong bonding that produces the most advantageous mechanical properties in whisker-reinforced composites has yet to be identified [42, 43]. Figure 2.5 shows the vapor infiltration process for casting of composites.

2.2.2.2 Casting methods

Literature provides extensive processing information, but the open literature typically lacks detailed information about commercial processes. Although liquid metal routes like stir casting and infiltration account for the biggest amount of primary production (about 65%) among commercial processes, their cheap cost makes them represent just about 27% of the composite market by value [44].

2.2.2.2.1 Stir casting

Typically, molten metal matrix and ceramic particle reinforcement are mixed together during the stir casting process. Particulates are dispersed and suspended in the molten metal through the use of high-energy mixing or another

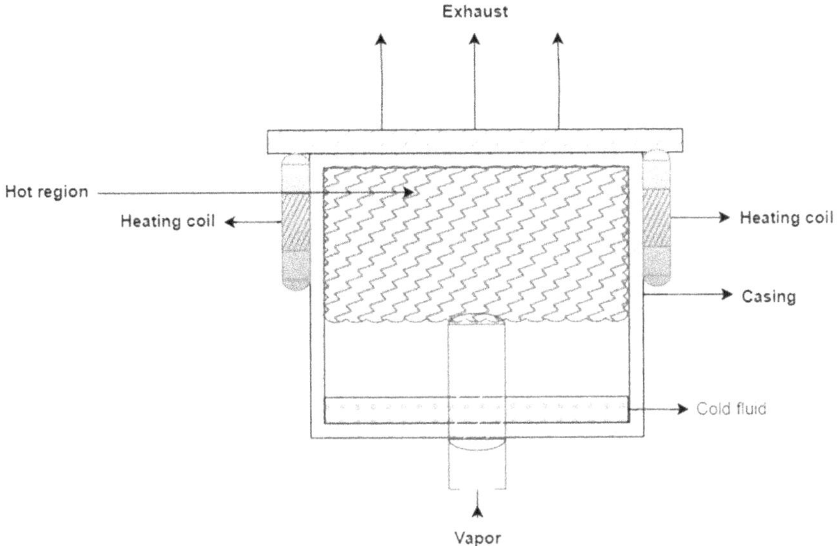

Figure 2.5 Vapor infiltration process.

suitable method. Usually, the specifics of these procedures are confidential knowledge. After that, the suspended slurry is cast as a rolling bloom, extrusion billet, or foundry ingot. The product is also marketed commercially as remelt stock, in which normal or modified metal casting techniques are used to create near-net form components. Size, density, and chemical reactivity of the reinforcement are crucial factors. This technique provides products ranging from 8% to 38%. Applications that need for cheap costs and large production numbers frequently use this technology [45]. Reddy et al. [46] fabricated Aluminium 6063 with TiC composite by using stir casting method as shown in Figure 2.6 and stated that the reinforcement of particles improved the mechanical properties like tensile strength and hardness. Rohatgi et al. [47] attempted to add SiC into A356 by using stir casting furnace and stated that addition of SiC could make parts cheaper and lighter.

2.2.2.3 Other methods

2.2.2.3.1 Deposition of material using spray process

Spray deposition is the method of creating matrix material into a sufficient dispersal of droplets by injecting heated reinforcing particles into pressure-controlled inert gas jets. In their evaluation and discussion of several synthesis methods for creating composites using particle technology, Srivatsan and Lavernia [48] found that spray deposition offers excellent opportunities for creating composites of high quality. A thorough analysis of the several kinds of composites production procedures, their mechanical characterization, and

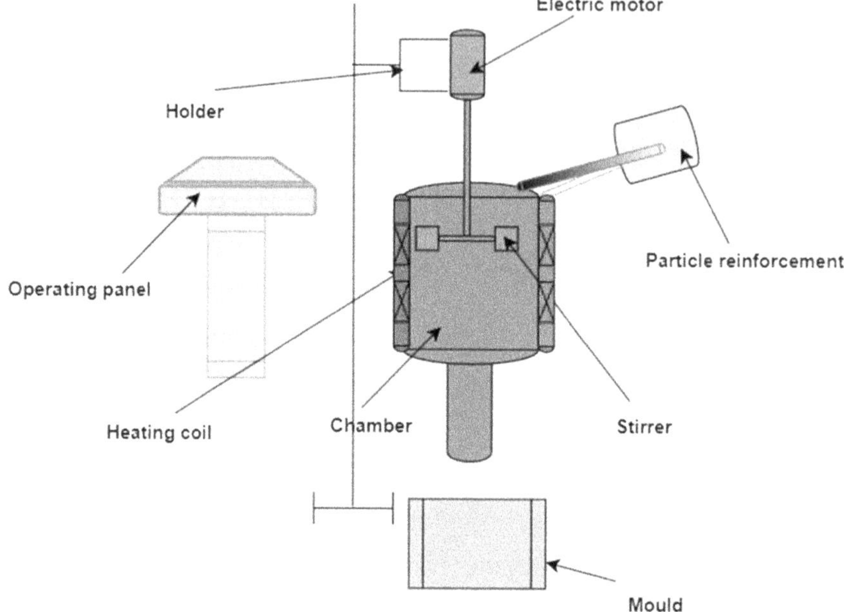

Figure 2.6 Stir casting machine for liquid processing route of composites preparation.

their use in many industries was provided by Mistry and Gohil [49]. They discovered that the spray deposition approach helps the MMC achieve the least amount of matrix material reaction with reinforcement by offering a higher production rate and a shorter solidification time.

2.3 CONCLUSIONS

Among the different methods, stir casting is the main, well-known, and cost-effective method for making composites. The most promising production techniques for making MMCs include semi-solid, PM, and stir/squeeze casting. The squeeze pressure is the primary parameter that affects the mechanical characteristics in the stir/squeeze casting process. A uniform distribution of the reinforcement particles, wettability, porosity, erosion of the stirrer blades, and reinforcement mixing rate are the main issues in the stir casting process. Apart from the sustainable development of MMCs through the use of waste reinforcement particles generated in industrial processes and recycled matrix, these problems pose promising avenues for future research. For the manufacturing of aluminium alloy based composites, reinforced with any type of reinforcing material, a bottom tapping stir casting furnace with squeezing attachment and preferably electromagnetic and ultrasonic stirring would be optimal. High-quality MMCs with improved mechanical properties can be made by carefully choosing the production method and factors with matrix, reinforcing material, wetting agent, and additives.

2.4 FUTURE ASPECTS FOR COMPOSITE DEVELOPMENT

The manufacture of MMCs is fraught with challenges, as previously indicated. There is a wide range in the properties of MMCs manufactured using various casting processes. Therefore, it may be challenging to choose an appropriate method for a certain application. In order to create various MMC compositions, it is important to optimize the process parameters and conditions. Process parameters for many newly introduced and existing reinforcing materials, particularly nanoparticles, cannot be finalized due to the lack of available research. Since porosity is an inevitable by-product of any casting process, strategies for mitigating it should be thoroughly investigated. A further parameter that pointedly disturbs the mechanical properties is the uniform distribution of the reinforcing materials. Similarly, the bonding between the reinforcement and matrix, which is controlled by wettability, also has a significant impact on the mechanical characteristics [2, 50, 51]. Very little research has been done on wettability other than that done by Hashim et al. [44]. Reusing materials is crucial as interest in creating environment-friendly methods grows; nevertheless, despite research suggesting otherwise, this issue has received little attention [7, 11]. When it comes to producing MMCs, using scrap, waste, or spent materials for both the matrix and the reinforcement is still a relatively unexplored field that offers plenty of opportunity for specialized research.

REFERENCES

1. S. Agrawal, K. Singh, and P. Sarkar, "Comparative investigation on the wear and friction behaviors of carbon fiber reinforced polymer composites under dry sliding, oil lubrication and inert gas environment," *Materials Today: Proceedings*, vol. 5, pp. 1250–1256, 2018.
2. P. Babalola, C. Bolu, A. Inegbenebor, and K. Odunfa, "Development of aluminium matrix composites: a review," *Online International Journal of Engineering and Technology Research*, vol. 2, pp. 1–11, 2014.
3. X. Dangsheng, "Friction and wear properties of UHMWPE composites reinforced with carbon fiber," *Materials Letters*, vol. 59, pp. 175–179, 2005.
4. L. Fang, Y. Leng, and P. Gao, "Processing and mechanical properties of HA/UHMWPE nanocomposites," *Biomaterials*, vol. 27, pp. 3701–3707, 2006.
5. S. Suresh, *Fundamentals of metal-matrix composites*: Elsevier, 2013.
6. M. Venkatesan, K. Palanikumar, and S. R. Boopathy, "Experimental investigation and analysis on the wear properties of glass fiber and CNT reinforced hybrid polymer composites," *Science and Engineering of Composite Materials*, vol. 25, pp. 963–974, 2018.

7. J. Joel and M. A. Xavior, "Aluminium alloy composites and its machinability studies: A review," *Materials Today: Proceedings*, vol. 5, pp. 13556–13562, 2018.

8. R. Rajan, P. Kah, B. Mvola, and J. Martikainen, "Trends in aluminium alloy development and their joining methods," *Reviews on Advanced Materials Science*, vol. 44, pp. 383–397, 2016.

9. A. K. Srivastava, S. Dwivedi, A. Nag, D. Kumar, A. R. Dixit, and S. Hloch, "Microstructural, mechanical and tribological performance of a magnesium alloy $AZ_{31}B/Si_3N_4$/eggshell surface composite produced by solid-state multi-pass friction stir processing," *Materials Chemistry and Physics*, vol. 301, p. 127694, 2023.

10. P. Ramesh, S. Vivekanandan, Sivaramakrishnan, and D. Prakash, "Performance optimization of an engine for canola oil blended diesel with Al_2O_3 nanoparticles through single and multi-objective optimization techniques," *Fuel*, vol. 288, p. 119617, 2021.

11. Ş. Karabulut, H. Karakoç, and R. Çıtak, "Influence of B4C particle reinforcement on mechanical and machining properties of Al6061/B4C composites," *Composites Part B: Engineering*, vol. 101, pp. 87–98, 2016.

12. P. Vasanthakumar, K. Sekar, and K. Venkatesh, "Recent developments in powder metallurgy based aluminium alloy composite for aerospace applications," *Materials Today: Proceedings*, vol. 18, pp. 5400–5409, 2019.

13. R. Sultana, U. Banik, P. K. Nandy, M. N. Huda, and M. Ismail, "Bio-oil production from rubber seed cake via pyrolysis: Process parameter optimization and physicochemical characterization," *Energy Conversion and Management: X*, vol. 20, p. 100429, 2023.

14. M. K. Surappa, "Aluminium matrix composites: Challenges and opportunities," *Sadhana*, vol. 28, pp. 319–334, 2003.

15. A. Macke, B. Schultz, and P. Rohatgi, "Metal matrix composites," *Advanced Materials and Processes*, vol. 170, pp. 19–23, 2012.

16. K. U. Kainer, "Basics of metal matrix composites," Metal matrix composites: Custom-made materials for automotive and aerospace engineering: Wiley, pp. 1–54, 2006.

17. C. C. Okpala, "Nanocomposites: An overview," *International Journal of Engineering Research and Development*, vol. 8, pp. 17–23, 2013.

18. R. A. Vaia and H. D. Wagner, "Framework for nanocomposites," *Materials Today*, vol. 7, pp. 32–37, 2004.

19. Z. Liu, S. Xu, B. Xiao, P. Xue, W. Wang, and Z. Ma, "Effect of ball-milling time on mechanical properties of carbon nanotubes reinforced aluminum matrix composites," *Composites Part A: Applied Science and Manufacturing*, vol. 43, pp. 2161–2168, 2012.

20. J. Mendoza-Duarte, I. Estrada-Guel, C. Carreño-Gallardo, and R. Martínez-Sánchez, "Study of Al composites prepared by high-energy ball milling: Effect of processing conditions," *Journal of Alloys and Compounds*, vol. 643, pp. S172–S177, 2015.

21. C. Suryanarayana and N. Al-Aqeeli, "Mechanically alloyed nanocomposites," *Progress in Materials Science*, vol. 58, pp. 383–502, 2013.

22. F. Hadef, "Effect of high-energy ball milling on structure and properties of some intermetallic alloys: A mini review," *Metallography, Microstructure, and Analysis*, vol. 8, pp. 430–444, 2019.

23. D. Zhang, "Processing of advanced materials using high-energy mechanical milling," *Progress in Materials Science*, vol. 49, pp. 537–560, 2004.

24. D. Agrawal, "Microwave sintering, brazing and melting of metallic materials," in *Sohn International Symposium; Advanced Processing of Metals and Materials Volume 4: New, Improved and Existing Technologies: Non-Ferrous Materials Extraction and Processing*, 2006, pp. 183–192.

25. R. Roy, D. Agrawal, J. Cheng, and S. Gedevanishvili, "Full sintering of powdered-metal bodies in a microwave field," *Nature*, vol. 399, pp. 668–670, 1999.

26. K. H. Brosnan, G. L. Messing, and D. K. Agrawal, "Microwave sintering of alumina at 2.45 GHz," *Journal of the American Ceramic Society*, vol. 86, pp. 1307–1312, 2003.

27. A. Chatterjee, T. Basak, and K. Ayappa, "Analysis of microwave sintering of ceramics," *AIChE Journal*, vol. 44, pp. 2302–2311, 1998.

28. J. Sethi, S. Das, and K. Das, "Evaluating the influence of milling time, and sintering temperature and time on the microstructural changes and mechanical properties of $Al-Y_2W_3O_{12}$-AlN hybrid composites," *Powder Technology*, vol. 377, pp. 244–256, 2021.

29. S. Takayama, Y. Saito, M. Sato, T. Nagasaka, T. Muroga, and Y. Ninomiya, "Sintering behavior of metal powders involving microwave-enhanced chemical reaction," *Japanese Journal of Applied Physics*, vol. 45, p. 1816, 2006.

30. M. N. Hassan, M. M. Mahmoud, G. Link, A. Abd El-Fattah, and S. Kandil, "Sintering of naturally derived hydroxyapatite using high frequency microwave processing," *Journal of Alloys and Compounds*, vol. 682, pp. 107–114, 2016.

31. W. E. Mahmoud, F. Al-Agel, and E. Al-Arfaj, "The influence of sintering temperature on the engineered nanoporous titania ceramics," *Materials Letters*, vol. 96, pp. 146–148, 2013.

32. B. P. Sahoo and D. Das, "Critical review on liquid state processing of aluminium based metal matrix nano-composites," *Materials Today: Proceedings*, vol. 19, pp. 493–500, 2019.

33. M. Aghajanian, M. Rocazella, J. T. Burke, and S. Keck, "The fabrication of metal matrix composites by a pressureless infiltration technique," *Journal of Materials Science*, vol. 26, pp. 447–454, 1991.

34. V. Michaud and A. Mortensen, "Infiltration processing of fibre reinforced composites: Governing phenomena," *Composites Part A: Applied Science and Manufacturing*, vol. 32, pp. 981–996, 2001.

35. K. Sree Manu, L. Ajay Raag, T. Rajan, M. Gupta, and B. Pai, "Liquid metal infiltration processing of metallic composites: A critical review," *Metallurgical and Materials Transactions B*, vol. 47, pp. 2799–2819, 2016.

36. A. Cook and P. Werner, "Pressure infiltration casting of metal matrix composites," *Materials Science and Engineering: A*, vol. 144, pp. 189–206, 1991.

37. R. Etemadi, B. Wang, K. Pillai, B. Niroumand, E. Omrani, and P. Rohatgi, "Pressure infiltration processes to synthesize metal matrix composites: A review of metal matrix composites, the technology and process simulation," *Materials and Manufacturing Processes*, vol. 33, pp. 1261–1290, 2018.

38. C. Guo, X. He, S. Ren, and X. Qu, "Effect of (0–40) wt.% Si addition to Al on the thermal conductivity and thermal expansion of diamond/Al composites by pressure infiltration," *Journal of Alloys and Compounds*, vol. 664, pp. 777–783, 2016.

39. J. Narciso, J. Molina, A. Rodríguez, F. Rodríguez-Reinoso, and E. Louis, "Effects of infiltration pressure on mechanical properties of Al-12Si/graphite composites for piston engines," *Composites Part B: Engineering*, vol. 91, pp. 441–447, 2016.

40. C. Marichy, V. Salles, X. Jaurand, A. Etiemble, T. Douillard, J. Faugier-Tovar *et al.*, "Fabrication of BN membranes containing high density of cylindrical pores using an elegant approach," *RSC Advances*, vol. 7, pp. 20709–20715, 2017.

41. X. Zhou, L. He, X. Cao, Z. Xu, R. Mu, J. Sun *et al.*, "$La_2(Zr_{0.7}Ce_{0.3})_2O_7$ thermal barrier coatings prepared by electron beam-physical vapor deposition that are resistant to high temperature attack by molten silicate," *Corrosion Science*, vol. 115, pp. 143–151, 2017.

42. W. F. Ingram and J. S. Jur, "Properties and applications of vapor infiltration into polymeric substrates," *JOM*, vol. 71, pp. 238–245, 2019.
43. N. H. Tai and T. W. Chou, "Analytical modeling of chemical vapor infiltration in fabrication of ceramic composites," *Journal of the American Ceramic Society*, vol. 72, pp. 414–420, 1989.
44. J. Hashim, L. Looney, and M. Hashmi, "Metal matrix composites: Production by the stir casting method," *Journal of Materials Processing Technology*, vol. 92, pp. 1–7, 1999.
45. S. Soltani, R. Azari Khosroshahi, R. Taherzadeh Mousavian, Z.-Y. Jiang, A. Fadavi Boostani, and D. Brabazon, "Stir casting process for manufacture of Al-SiC composites," *Rare Metals*, vol. 36, pp. 581–590, 2017.
46. P. V. Reddy, P. R. Prasad, D. M. Krishnudu, and E. V. Goud, "An investigation on mechanical and wear characteristics of Al 6063/TiC metal matrix composites using RSM," *Journal of Bio- and Tribo-corrosion*, vol. 5, p. 90, 2019.
47. P. Rohatgi, K. Pasciak, C. Narendranath, S. Ray, and A. Sachdev, "Evolution of microstructure and local thermal conditions during directional solidification of A356-SiC particle composites," *Journal of Materials Science*, vol. 29, pp. 5357–5366, 1994.
48. T. S. Srivatsan and E. Lavernia, "Use of spray techniques to synthesize particulate-reinforced metal-matrix composites," *Journal of Materials Science*, vol. 27, pp. 5965–5981, 1992.
49. J. M. Mistry and P. P. Gohil, "Research review of diversified reinforcement on aluminum metal matrix composites: Fabrication processes and mechanical characterization," *Science and Engineering of Composite Materials*, vol. 25, pp. 633–647, 2018.
50. B. P. Chang, H. M. Akil, M. G. Affendy, A. Khan, and R. B. M. Nasir, "Comparative study of wear performance of particulate and fiber-reinforced nano-ZnO/ultra-high molecular weight polyethylene hybrid composites using response surface methodology," *Materials & Design*, vol. 63, pp. 805–819, 2014.
51. W. Chen, F. Li, G. Han, J. Xia, L. Wang, J. Tu *et al.*, "Tribological behavior of carbon-nanotube-filled PTFE composites," *Tribology Letters*, vol. 15, pp. 275–278, 2003.

Chapter 3

Recent manufacturing approaches for composite materials

E. Hemachandran, Pavitra Singh, and Vijay K. Singh

3.1 INTRODUCTION

In the heart of today's industrial landscape, a profound transformation is driven by the relentless pursuit of innovation and efficiency. Composites with their unique blend of strength, lightness, and versatility have captured the imagination of industries ranging from aerospace to automotive, from construction to renewable energy [1]. As the demands for high-performance, sustainable materials intensify, composite materials have emerged as a beacon of possibility, offering a synthesis of properties that traditional materials struggle to match [2, 3]. However, unlocking the full potential of composites requires more than just their inherent properties; it demands a precise fusion of cutting-edge manufacturing techniques and engineering prowess.

Additive manufacturing (AM), often heralded as the cornerstone of the next industrial revolution, holds the promise of intricate designs and customized structures previously unimaginable [2, 4]. In the world of composites, AM opens doors to complex geometries, enabling the creation of lightweight yet robust components for everything from aircraft fuselages to medical implants. Beyond additive techniques, automation stands tall as a game changer, streamlining production processes and ensuring consistency and precision in every composite layer laid. Robots, with their unwavering accuracy and tireless efficiency, are becoming the artisans of the composite world, weaving carbon fibers into intricate patterns with unparalleled dexterity [2].

The fusion of automation and composites transcends traditional manufacturing constraints, allowing for rapid prototyping, reduced waste, and enhanced scalability [5]. Moreover, as industries race toward sustainability, advanced manufacturing offers a pathway to greener practices within the realm of composite materials. Recycling and reusing composite components, once a formidable challenge, are now made feasible through innovative manufacturing technologies, paving the way for a circular economy. This introduction sets the stage for a comprehensive exploration of how these advanced techniques are not mere tools but catalysts for progress in the composite materials industry.

DOI: 10.1201/9781003564355-3

The dream of electric aviation, once seen as distant, is brought closer to reality as composite structures, crafted with precision, pave the way for sustainable air travel. Meanwhile, the automotive industry is undergoing a metamorphosis, shedding its reliance on traditional metals in favor of composites that promise unparalleled strength-to-weight ratios. Electric vehicles, with their quest for extended range and enhanced safety, find a natural ally in composite materials, which made them all the more accessible through advanced manufacturing. The construction sector, too, experiences a renaissance as architects and engineers embrace composites for their durability, corrosion resistance, and design flexibility. With techniques like 3D printing of concrete composites, buildings rise from the ground with unprecedented speed and elegance, redefining skylines and urban landscapes [3]. Renewable energy, a cornerstone of our sustainable future, finds a formidable ally in composite materials engineered through advanced manufacturing. Wind turbine blades, vast and majestic, harness the power of the wind with lightweight yet sturdy composite constructions, maximizing energy output while minimizing environmental impact. The marine industry, sailing toward greener horizons, embraces composite materials for hulls and structures that defy corrosion and offer unmatched longevity.

Advanced manufacturing techniques, from 3D printing, fused filament fabrication (FFF), and stereolithography, lie at the heart of these innovations, pushing the boundaries of what is possible with composites [6]. As we delve deeper into this exploration, it becomes evident that the synergy between advanced manufacturing and composite materials is not just about efficiency—it is about unlocking new frontiers of design. Architects sculpt buildings with curves that defy gravity, made possible by composites shaped with precision by robotic arms guided by intricate algorithms [7]. Automotive designers reimagine vehicles, shedding the constraints of traditional forms to create aerodynamic marvels that slice through the air with minimal resistance.

The journey of a composite component, from its digital design to its physical manifestation, is a balance of technology and innovation supported by advanced manufacturing. Yet, amidst this wave of progress, challenges persist—challenges that demand not just technical expertise but an integral approach to sustainability, cost-effectiveness, and scalability. Material waste, a concern in any manufacturing process, is tackled through novel recycling methods that reclaim and repurpose composite scraps into new, usable materials. Scalability, crucial for widespread adoption, sees advancements as manufacturing techniques evolve to meet the demands of mass production without compromising quality. Cost-effectiveness, often a barrier to entry for innovative materials, is addressed through efficiencies gained in production, making composites increasingly competitive with traditional counterparts. Advanced manufacturing methods for composites involve various techniques that enhance efficiency, precision, and customization. Some of the notable methods are presented in this chapter.

3.2 3D PRINTING

Different methods of AM include material extrusion, VAT polymerization, powder bed fusion, material jetting, binder jetting, and 3D plotting. The fabrication and performance of parts through AM can vary depending on the mechanisms and scopes of these various processes. Three-dimensional printing has emerged as a hot topic in material fabrication and has captured the attention of researchers in the field of materials science [3, 6]. This technology has seen substantial expansion recently and is anticipated to radically transform manufacturing industries by enabling the production of advanced, high-performance materials for future applications [8, 9].

Three-dimensional printing has emerged as a significant technology for composite manufacturing [10]. Several methods within 3D printing can be utilized for composites, allowing for complex geometries, lightweight structures, and tailored material properties [11]. Decreased material wastage, flexible design options, and the ability to produce intricate structures have driven the integration of 3D printing technology across automotive, aviation, medical, and various engineering fields [12, 13]. Moreover, advancements in 3D printing research have opened significant avenues for scientists to craft complex structures with specific properties. AM techniques, such as FFF or SLA, can be used to create composite parts layer by layer. This method allows for complex geometries and customization without the need for traditional molds. Here we explore the advanced manufacturing methods for composite materials, illuminating how these techniques reshape entire sectors and propelling us into a future where ingenuity knows no bounds. The AM process involves a series of steps that begin with the initial concept and extend through to the production of the final item. The specifics of these steps can vary based on the selected production technology. Yet, these essential phases are consistent across various manufacturing methods, suitable for creating both prototype models and functional end-use products. Digital model creation process starts with the generation of a digital version of the object through CAD (computer-aided design) software. Alternative methods, such as reverse engineering and 3D scanning, may also be utilized to produce this digital representation. STL (Standard Triangle Language) file conversion in the subsequent phase involves converting the digital model into an STL format. This format is essential for representing the model's surface geometry. Slicing the model, now in STL format, is then processed by slicing software. The slicing stage is critical for determining the quality of the finished print. The software processes the STL file to produce G-codes, similar to those used in CNC (computer numerical control) machining. These codes accurately control the extruder and the platform's movements during the print job. Preparing for printing with the STL model converted into G-codes implies the printer is ready to start the actual printing. The approach to printing can vary depending on the specific AM technology in use. For FDM (fused deposition modeling), the print head is programmed to deposit melted material layer by layer, following the instructions in the G-code. The G-code carefully dictates the path of the print head, the volume of material being deposited, and the duration of extrusion.

3.2.1 Continuous fiber 3D printing

Continuous fiber 3D printing method involves depositing a thermoplastic material, such as nylon, along with continuous fibers (such as carbon or glass) in a continuous manner. This allows for the creation of parts with excellent strength-to-weight ratios and directional reinforcement. Utilizing 3D printing for the production of continuous fiber-reinforced plastics (CFRPs) offers benefits such as the ability to manufacture intricate shapes quickly. The mechanical properties of CFRPs, which are influenced by the orientation of the continuous fibers, dictate the strength of the printed material.

The process starts with the deposition of a thermoplastic matrix material, often in filament form, through a heated nozzle. This material acts as the binding agent for the continuous fibers. Simultaneously, continuous fibers are precisely laid down onto the part being built. These fibers are often spooled through the print head, allowing for continuous placement along the path of the print. The 3D printer builds the part layer by layer, with the thermoplastic material serving as the matrix to hold the continuous fibers in place. The fibers are typically oriented in specific directions based on the design requirements. As each layer is deposited, heat or pressure is applied to fuse the thermoplastic material and the continuous fibers. This fusion creates a strong bond between the fibers and the matrix, resulting in a robust composite structure. After each layer is deposited and fused, it is allowed to cool and solidify before the next layer is added on top. This ensures that the part retains its shape and structural integrity. Depending on the specific requirements of the part, post-processing steps such as curing in an oven, additional heating, or surface finishing may be performed. The different methods for 3D printing CFRPs present varied approaches to integrating continuous fibers into the printing process [8]. Two prominent approaches to integrate continuous fibers with thermoplastic matrix materials are out-of-nozzle impregnation and in-nozzle impregnation, as shown in Figure 3.1.

Out-of-nozzle impregnation technique involves combining continuous fibers with the thermoplastic matrix material outside of the 3D printer's nozzle. Typically, the fibers are pre-impregnated with resin. These resin-impregnated fibers are then introduced into the printer alongside the thermoplastic filament. Upon passing through the heated nozzle, the fibers and filament are heated and fused, resulting in a composite material. This method provides precise control over fiber orientation and placement but may require specialized equipment for pre-impregnating the fibers. In-nozzle impregnation integrates the continuous fibers with the thermoplastic filament within the printer's nozzle. The process starts with both the fibers and thermoplastic filament separately fed into the nozzle. As the fibers pass through the

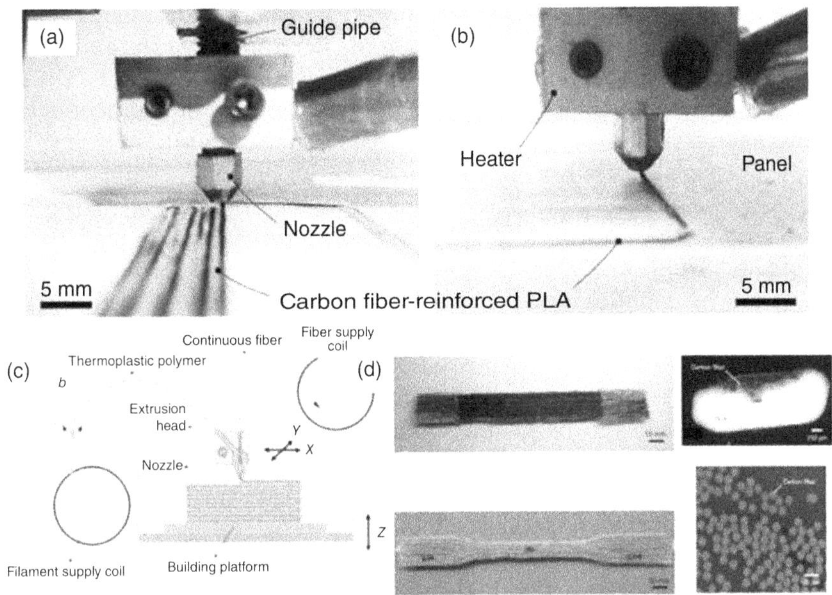

Figure 3.1 Continuous fibers with thermoplastic matrix materials are out-of-nozzle impregnation and in-nozzle impregnation.

heated nozzle, they are impregnated with the molten thermoplastic material. The combined material is then extruded onto the build platform or previous layers to create the composite part. This technique simplifies the process by eliminating the need for pre-impregnated fibers but may have limitations in precise fiber orientation control compared to out-of-nozzle impregnation. Printing with semifinished CFRP filaments involves using semifinished CFRP filaments directly in the 3D printing process. These filaments come ready-made with continuous fibers already impregnated with thermoplastic resin. The filaments are fed into the printer like conventional thermoplastic filaments. As they pass through the nozzle and are heated, the resin matrix melts, binding the continuous fibers together to form the composite material. This approach offers convenience as the filaments come pre-made with the desired fiber orientation and resin matrix, simplifying the printing process and ensuring consistent quality in the composite parts. However, it may have limitations in flexibility as the fiber orientation and resin matrix are predetermined by the semifinished filaments [6]. Continuous fiber 3D printing for composites offers a multitude of advantages. First, it provides an enhanced strength-to-weight ratio, harnessing continuous fibers to produce composite parts with exceptional strength and stiffness, making them perfect for applications prioritizing lightweight design without compromising structural integrity. Second, the method allows for customized fiber orientation, granting designers precise control over how continuous fibers are arranged within the part. This customization capability enables tailoring mechanical properties like tensile strength, flexural strength, and impact resistance to exact specifications. Third, the technology facilitates the creation of intricate geometries that would be challenging or impossible with traditional methods, including internal structures, hollow sections, and complex designs. Moreover, continuous fiber 3D printing minimizes material waste due to its additive nature, depositing material only where needed, thus significantly reducing waste compared to subtractive manufacturing techniques. Additionally, the process streamlines efficient prototyping and production by enabling rapid iteration and design testing. It also supports cost-effective on-demand production of customized composite parts, eliminating the requirement for expensive tooling or molds. Furthermore, the use of continuous fibers enhances fatigue resistance by distributing loads more evenly throughout the part, improving durability, which is particularly advantageous in dynamic applications. Finally, the method offers material versatility, accommodating various continuous fibers such as carbon, glass, and aramid, along with a range of thermoplastic matrix materials. This versatility provides a wide spectrum of material properties and performance characteristics to meet diverse and specific requirements in composite manufacturing [25]. Continuous fiber 3D printing is widely employed across industries for a range of applications. In aerospace, this technology produces lightweight yet robust composite parts vital for aircraft, UAVs (unmanned aerial vehicles), satellites, and spacecraft, contributing to improved fuel efficiency and performance [15]. Automotive sectors utilize it to manufacture structural components, body panels, and lightweight parts, particularly beneficial for electric vehicles seeking enhanced range and efficiency. In the medical field, customization capabilities enable the creation of patient-specific implants, prosthetics, and medical devices, ensuring optimal fit and functionality. Sporting goods such as high-performance bicycle frames, tennis rackets, and helmets benefit from the strength-to-weight advantages, providing athletes with superior performance and durability. Additionally, in the marine industry, continuous fiber 3D printing is employed to fabricate boat hulls, propellers, and marine components, offering enhanced strength and corrosion resistance for marine applications. Continuous fiber 3D printing is an exciting advancement in composite manufacturing, offering a combination of strength, customization, and design freedom that opens up new possibilities across various industries.

3.2.2 Fused filament fabrication (FFF)

FFF is a common 3D printing method where a thermoplastic filament is heated and extruded through a nozzle layer by layer to create a part. In the case of composite 3D printing, the filament can be infused with chopped fibers, such as carbon or glass, to enhance the mechanical properties of the printed part [5].

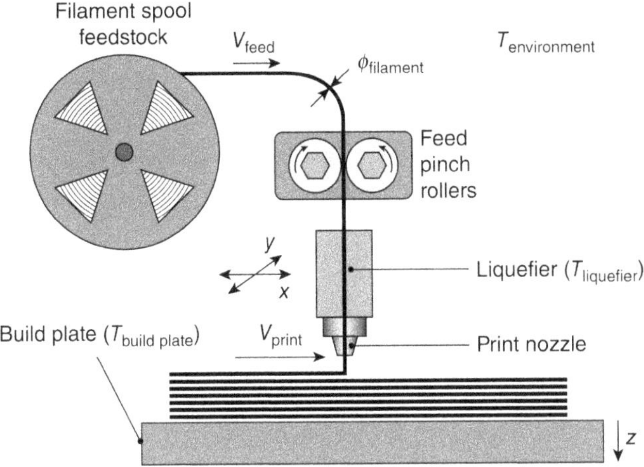

Figure 3.2 Schematic image of thermoplastic filament spool with a print nozzle for 3D object creation.

FFF, also known as FDM, is a widely used 3D printing method for composite manufacturing [17]. In this process, a thermoplastic filament is heated and extruded through a nozzle, layer by layer, to create a 3D object (Figure 3.2).

In FFF, the material is extruded in continuous, single beads along the x–y plane. These beads have specific thicknesses and cross-sectional shapes determined by the FFF process settings and the design of the extruder head. As the beads are deposited, they adhere to the surrounding beads, building up a layer based on their solidification characteristics. After one layer is completed, the printer adjusts the z-height, and this layering process continues, stacking one layer on top of another. This method of bead-to-bead adhesion (in the x–y, x–z, and y–z directions) significantly affects the dimensional precision and mechanical properties of the printed objects. When used for composites, this method involves adding reinforcing fibers to the thermoplastic filament. Here is an overview of FFF for composite manufacturing.

The process of FFF for composites begins with the preparation of a composite filament, which includes a thermoplastic matrix material (like ABS, PLA, PETG) blended with chopped fibers such as carbon, glass, or aramid (Figure 3.3).

Varying percentages of fibers are mixed to achieve specific properties. Next, the composite filament is fed into a heated nozzle of the 3D printer where it melts the thermoplastic material. The molten composite material is then extruded onto the build platform in a process known as heating and extrusion. The printer head moves along a predetermined path, depositing the molten composite material layer by layer, allowing each layer to bond with the previous one, gradually constructing the 3D object. While the fibers within the filament are initially randomly oriented, their direction can be somewhat

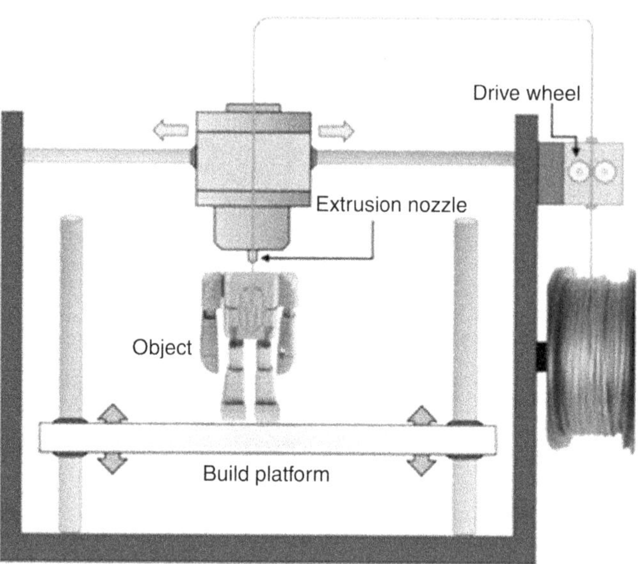

Figure 3.3 The process of FFF for composites.

controlled by adjusting printing parameters such as nozzle temperature, print speed, and layer thickness. After each layer is deposited, it quickly cools and solidifies, crucial for maintaining the shape and structural integrity of the part. For complex geometries or overhangs, support structures may be incorporated during printing using the same composite material, which can be removed once printing is finished, allowing for intricate designs to be realized.

In the FDM technique, polymers are successively extruded and stacked to craft a product. This AM method is distinguished by its ability to create polymers that are structurally strong, boast a superior surface quality, and exhibit longevity, all at an affordable cost. The raw material for FDM, usually presented as long filament strands wound around a spool, has a diameter ranging from 1.8 mm to 3 mm, depending on the nozzle size of the machine. Initially, this filament is fed into a melt head that heats it above its glass transition temperature, transforming it into a flowable melt. This molten polymer is then directed through a nozzle and extruded, with the nozzle moving across the XY axis as directed by the system's design inputs. FDM 3D printing configuration procedure entails the layer-by-layer extrusion of the molten filament onto the construction platform, progressively building up the design. After finishing a layer, the platform descends in the Z-axis to accommodate the addition of the next layer, which adheres to the preceding one. The thickness of each layer, and thereby the print's resolution, is determined by the platform's movement, with the capability of achieving layer heights as fine as 20 μm for enhanced detail. The integrity of layer bonding is pivotal in determining the mechanical strength of parts or composites produced through AM.

In the printing process, the layers merge without the need for any external force or pressure, primarily due to the high temperatures of the freshly deposited layer [19]. The phenomenon of interlayer bonding in FDM has been emphasized in several studies, highlighting how adjacent layers undergo a form of local welding. This interlayer fusion is critical because the overall strength of the 3D printed object is greatly influenced by the meso-structural characteristics and the bonding quality between layers. When subjected to stress, the first point of failure typically occurs at the site of weakest bonding. The mesostructure, in this context, refers to the development of necking between layers and the emergence of bonds and voids within these interstices.

Understanding how fibers are aligned within a composite is crucial for grasping composite properties. The orientation of fibers, induced by the flow of material, impacts the mechanical and thermal characteristics of a polypropylene and carbon fiber composite [20]. Their morphological studies indicated a tendency for fibers to align in the direction of printing. Interactions both among fibers and between fibers and the matrix were observed to improve in the composite. As a new layer was added atop a previously deposited one, the surface of the underlying layer began to remelt, fostering a connection between fibers across layers. This led to a composite with fibers oriented both longitudinally and orthogonally, interconnected throughout. Layer bonding mechanism in thermoplastic, identifying four key stages in the process, is as follows [21]. Initially, the interface is heated, enhancing the mobility and interaction of polymer molecules. Next, a close physical proximity between the adhering surfaces that are bonding is achieved. This is followed by the diffusion of molecules across the interface. Finally, the interface is cooled down to a temperature beneath the glass transition threshold.

FFF for composites offers a multitude of advantages in manufacturing. First, it enables cost-effective production, making it a viable option for small-scale manufacturing and prototyping of composite parts without the need for expensive tools or molds. Second, it provides customization and design freedom, allowing designers to create intricate geometries and custom shapes that might be challenging or costly using traditional methods. This flexibility is particularly beneficial for rapid prototyping and iterative design processes. Third, FFF enhances the strength, stiffness, and impact resistance of parts through the addition of continuous or chopped fibers, ensuring reinforced strength essential for applications requiring structural integrity. Additionally, the process results in reduced material waste as it is an AM method, depositing material only where needed, thus minimizing waste compared to subtractive methods. Moreover, FFF offers versatility in material choice, with a wide range of thermoplastic matrix materials and reinforcing fibers available. This allows for tailoring of mechanical and thermal properties to meet specific application requirements. Finally, the composites produced with FFF are inherently lightweight, making them ideal for weight-sensitive applications in aerospace, automotive, and sporting goods where weight reduction is critical [15].

FFF for composites finds diverse applications across industries. First, it is utilized for prototyping, enabling rapid creation of composite parts for design validation and testing purposes. Second, FFF is employed to manufacture custom components, producing tailored composite parts designed for specific applications in the aerospace, automotive, medical, and consumer goods sectors. Additionally, it facilitates the production of functional parts with reinforced strength and durability, suitable for end-use applications. Furthermore, the technology is utilized to create jigs, fixtures, and tooling, providing custom solutions for manufacturing processes. Overall, FFF for composites stands as a cost-effective, versatile, and accessible method for producing strong, lightweight parts with properties tailored to the application. Its applications continue to expand across various industries seeking the advantages of composite materials.

3.2.3 Carbon fiber-reinforced plastics (CFRP) with 3D printing

Carbon fiber-reinforced plastics with 3D printing combines traditional CFRP manufacturing techniques with 3D printing. This process involves the deposition of thermoplastic or thermoset materials infused with carbon fibers to form complex, high-performance structures. Continuous carbon fibers are laid down in the desired pattern, and then a polymer resin is added either through traditional methods or through 3D printing to create the final composite part. The process of carbon fiber-reinforced polymer with 3D printing begins with the preparation of a composite filament, comprising a thermoplastic or thermoset matrix material infused with chopped or continuous carbon fibers. These fibers are meticulously distributed throughout the filament to offer strength and reinforcement (Figure 3.4). Subsequently, the composite filament is loaded into a 3D printer, often utilizing an FFF or FDM technique [17].

Through this, the filament is heated and extruded via a nozzle, enabling it to be deposited layer by layer, gradually constructing the desired shape of the part. Notably, the printing process allows for controlled fiber orientation,

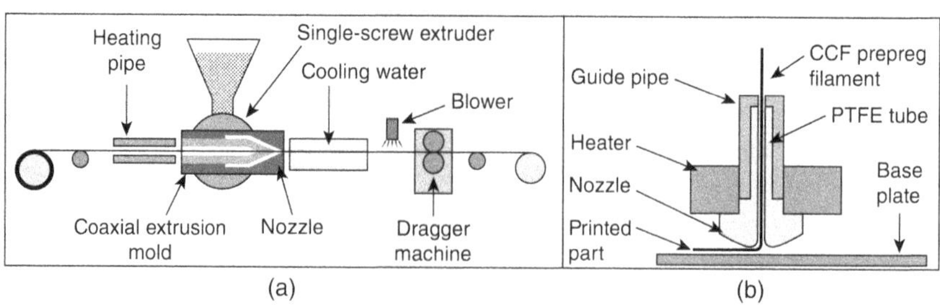

Figure 3.4 Single screw extruder with accessories. (a) Horizontal (b) Vertical.

where the direction of carbon fibers can be precisely adjusted. This capability enables the tailoring of mechanical properties, such as heightened stiffness along specific directions or augmented impact resistance in critical areas. The 3D printer continues this layer-by-layer building process, with each layer bonding seamlessly to the previous one, facilitating the creation of intricate geometries and internal structures within the part. Depending on the matrix material used, the part may undergo a consolidation and curing stage post-printing. This could involve heating the part to bond the layers or exposing it to ultraviolet light for resin curing in methods like SLA or digital light processing (DLP). Finally, after the part is fully printed and cured, it may undergo post-processing steps such as sanding, polishing, or coating to achieve the desired surface finish and properties required for the application. The advantages of CFRP with 3D printing are manifold. First, it offers an enhanced strength-to-weight ratio, harnessing the strength and stiffness of carbon fibers to provide robustness while reducing weight, a critical benefit in aerospace and automotive industries where weight savings are paramount. Second, 3D printing enables enhanced customization by allowing the production of intricate and customized geometries that would be challenging with traditional manufacturing methods, opening new design possibilities. Third, the process leads to minimized waste as AM, such as 3D printing, ensures material is deposited precisely where needed, reducing waste compared to subtractive techniques. Additionally, CFRP components exhibit improved performance characteristics such as heightened fatigue resistance, corrosion resistance, and dimensional stability. Fourth, 3D printing offers cost-effective prototyping, enabling swift prototyping and design iteration, thereby reducing lead times and costs associated with traditional prototyping. Furthermore, versatility is a key advantage, as a variety of carbon fiber types and matrix materials can be utilized to tailor the composite's properties to precise specifications, such as enhanced heat resistance or electrical conductivity. Finally, functional integration is enhanced through 3D printing, allowing for the seamless integration of features like internal channels, mounting points, and lattice structures directly into the design of the part, thereby improving functionality and efficiency.

The applications of carbon fiber reinforced polymer with 3D printing span various industries. In aerospace, this technology is integral for producing structural components, interior parts, and brackets crucial for aircraft and spacecraft. Automotive manufacturers leverage CFRP with 3D printing to create lightweight body panels, interior components, and specialized parts for racing cars, enhancing performance and efficiency. Within the medical field, CFRP with 3D printing enables the fabrication of custom prosthetics, implants tailored to individual patient needs, and precise surgical tools, revolutionizing patient care. Sporting goods like bicycle frames, tennis rackets, golf club shafts, and helmets are also manufactured with this technology for their lightweight and durable properties, benefiting athletes worldwide. In industrial settings, CFRP with 3D printing is used to fabricate custom tooling, jigs, and fixtures that streamline manufacturing processes and improve over-all efficiency. The field of CFRP with 3D printing continues to evolve rapidly, offering immense potential for innovation and efficiency in composite manu-facturing. Its capacity to produce lightweight, strong, and customizable parts positions it as a valuable technology with wide-ranging applications across diverse industries.

3.3 STEREOLITHOGRAPHY (SLA)

SLA is a resin-based 3D printing technology that can be utilized for advanced composites manufacturing by incorporating reinforcing materials such as car-bon fibers, glass fibers, or aramid fibers into the resin. This process results in composite parts with enhanced mechanical properties such as strength, stiff-ness, and durability [17].

The process of SLA for advanced composites manufacturing begins with the formulation of composite resins, achieved by blending a photopolymer resin with reinforcing fibers such as carbon, glass, or aramid. These fibers, whether in chopped, continuous, or fabric form, are uniformly dispersed within the resin to enhance its strength. Next, a 3D model of the desired com-ponent is crafted using CAD software and meticulously sliced into thin cross-sectional layers to guide the SLA printer. Upon setting up the SLA printer, the composite resin is carefully poured into the resin vat, where a build platform is submerged. Using a laser or UV light source, the resin is selectively cured in the specified pattern. Layer-by-layer printing ensues as the laser solidifies the resin wherever it strikes, with the build platform gradually descending to construct the entire part. During this process, controlled fiber orientation occurs, aligning the reinforcing fibers as per design requirements to optimize mechanical properties such as tensile strength and stiffness. Once printing is finished, the part is removed from the build platform for post-processing. This may involve washing to remove excess resin and curing for full hard-ening. Additionally, surface finishing techniques like sanding, polishing, or coating can be applied to meet specific requirements for the final product.

The manufacturing process of UV-cured composite materials is a meticu-lous and efficient method for producing sturdy, lightweight, and durable com-ponents (Figure 3.5).

This process unfolds in a series of steps, starting with the careful selec-tion of a UV-curable resin suitable for the intended application. Composed of monomers, oligomers, and photoinitiators, these resins are combined with reinforcing fibers like carbon, glass, or aramid fibers to create the base mate-rial. Thorough mixing ensures an even distribution of fibers within the resin matrix, often incorporating additives for specific properties such as flame resistance or color. Mold preparation follows, where the mold is designed to match the final part's shape and dimensions, with materials ranging from metals to composites. Surface preparation, including polishing or applying

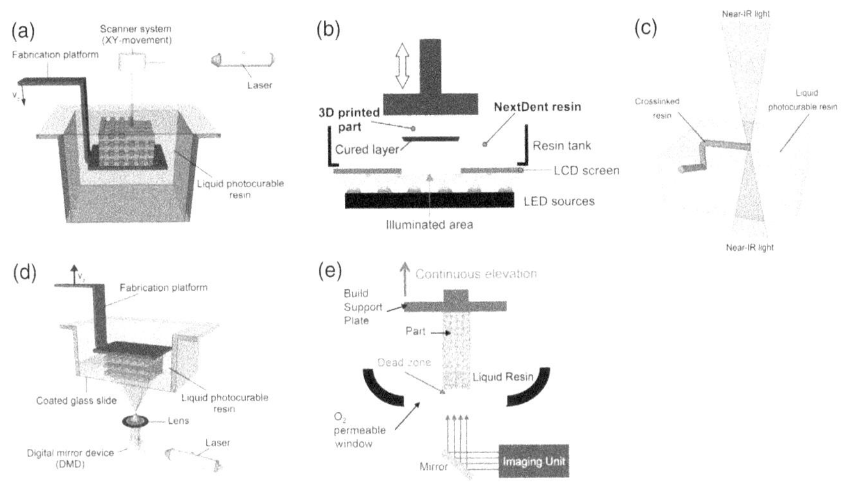

Figure 3.5 Schematic process for UV-cured composite materials.

a release agent, aids in easy part removal. The composite layup involves layering the resin and reinforcement mixture, known as the prepreg, into the mold, meticulously aligning fibers for desired orientation and thickness. Pressure, via vacuum bagging or pressing, consolidates the layers, removing excess resin and ensuring a strong bond. Curing under UV light activates the photoinitiators, prompting rapid polymerization, transforming the resin from liquid to solid in minutes. This process is closely controlled, allowing for adjustments in light intensity and exposure time. After curing, the part cools within the mold to stabilize dimensions, then delicately demolded with the aid of a release agent. Trimming and finishing follow, where excess material is removed, and final touches such as sanding, polishing, or coatings are applied for the desired texture and appearance. The advantages of UV-cured composites include rapid production cycles, precise control over curing, low energy consumption, and reduced waste and emissions. This method offers a structured approach to create high-performance parts tailored to specific requirements, where each stage from material selection to finishing plays a crucial role in achieving strength, durability, and efficiency in composite components.

SLA for advanced composites manufacturing offers a range of advantages, including enhanced mechanical properties with reinforcing fibers that significantly improve the strength, stiffness, and toughness of composite parts, akin to traditional composites but with the benefits of 3D printing. Designers benefit from the freedom and complexity to create intricate internal structures and complex geometries, optimizing designs for specific applications while reducing weight and ensuring structural integrity. With high-resolution capabilities, SLA 3D printers produce parts with fine details and smooth surface finishes crucial for tight tolerances and intricate features in composite components. Additionally, the process is efficient in material usage, minimizing waste as the resin is only cured where needed, contrasting with subtractive manufacturing. SLA also enables customization and rapid prototyping, allowing for quick iteration of designs and testing of various composite formulations and fiber orientations without the need for costly tooling. The versatility extends to the choice of reinforcing fibers such as carbon, glass, or aramid, providing tailored mechanical properties, and making SLA a viable option across diverse industries and applications.

3.4 DIRECT ENERGY DEPOSITION (DED) FOR COMPOSITE MANUFACTURING

DED is an AM method used in composite manufacturing to create complex geometries with precise control over material placement. Different classifications of DED systems are shown in Figure 3.6. When applied to composites, DED involves the use of a focused thermal energy source to melt and fuse both a matrix material and reinforcing fibers, such as continuous fibers or chopped fibers [24–26].

DED for composites involves several key steps in the manufacturing process. First, the composite feedstock is prepared, comprising a matrix material (typically a metal or polymer) combined with reinforcing fibers like carbon, glass, or aramid fibers. These fibers can be continuous strands or chopped fibers mixed into the matrix material. Next, a high-energy heat source, such as a laser or electron beam, is utilized to melt and fuse the composite feedstock material. This heat source is precisely directed onto the substrate or the previous layer of the part being constructed. The composite feedstock material is then fed into the path of the thermal energy source. As the material is heated, it melts and is deposited onto the substrate or the preceding

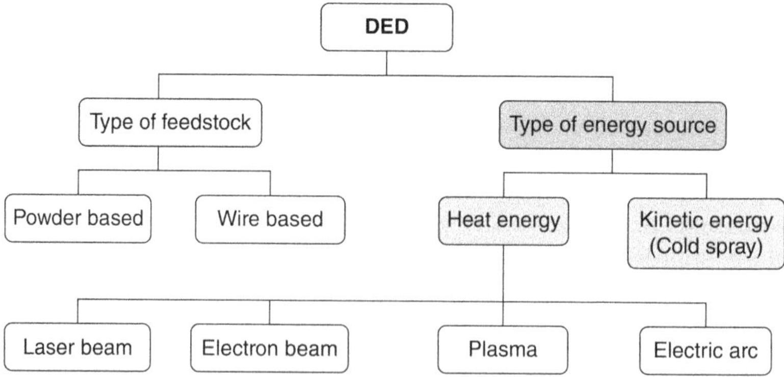

Figure 3.6 Classification of DED for composite manufacturing.

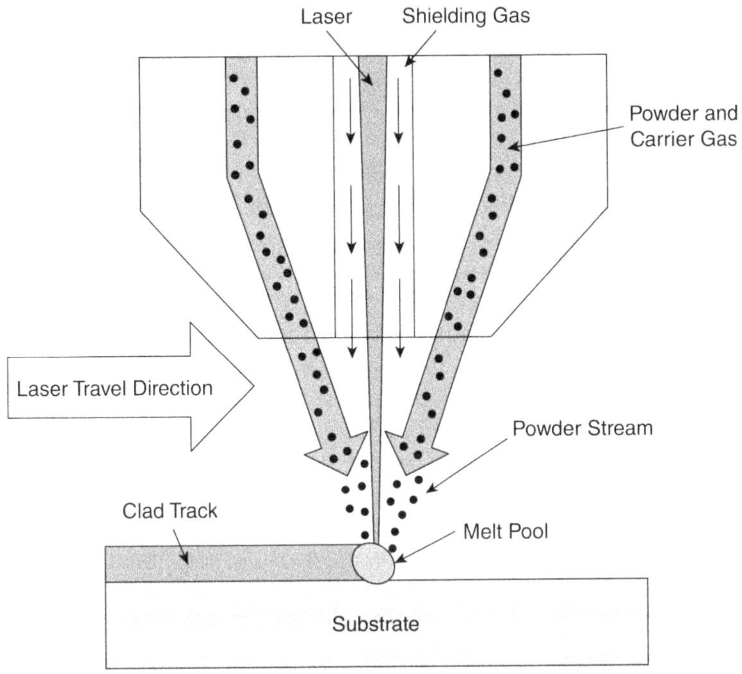

Figure 3.7 Process representing composite feedstock material into the path of thermal energy storage.

layer (Figure 3.7). This process continues, building up the part layer by layer following a predetermined toolpath created for the desired shape.

During deposition, the orientation and placement of the reinforcing fibers are precisely controlled, allowing for tailored mechanical properties and reinforcement in specific directions. As each layer is deposited and fused, it rapidly cools and solidifies, crucial for maintaining the shape and integrity of the part. Post-processing steps such as machining, surface finishing, or heat treatment may then be carried out to achieve the final desired properties of the composite part.

DED for composites offers several notable advantages. One key advantage is the ability to tailor material properties precisely. DED allows for meticulous control over the placement and orientation of reinforcing fibers within the composite part, resulting in customized mechanical properties such as strength, stiffness, and impact resistance. Additionally, the additive nature of DED facilitates the production of complex and intricate geometries that would be challenging or impossible with traditional manufacturing methods. Another benefit is the high deposition rates achievable with DED, making it suitable for the rapid manufacturing of large parts. This efficiency also leads to reduced waste since the material is added only where needed, making DED a relatively efficient process. Moreover, DED is versatile in its material capabilities and capable of working with metals, polymers, and ceramics. This versatility allows for the creation of hybrid composite parts with varying material properties within a single structure. Furthermore, DED's application extends to the repair and refurbishment of existing parts, where material is added precisely to the areas needing repair. Finally, DED can be used for in situ manufacturing and repair, particularly valuable in industries like aerospace, where on-site production or repair is often required.

DED for composites finds diverse applications across industries. In aerospace, it is utilized for engine components, turbine blades, and structural parts requiring both high strength and intricate geometries. Automotive sectors benefit from DED for manufacturing lightweight chassis parts, engine components, and brackets. Marine applications include propellers, hull structures, and marine components that demand corrosion resistance and strength. DED is also employed in the production of tooling and molds, providing enhanced durability and wear resistance. In the energy sector, it contributes to the creation of components for renewable energy systems such as wind turbine blades and solar panel supports. Overall, DED for composites stands as a versatile and advanced manufacturing method, offering the unique capability to produce high-performance parts with tailored properties. Its value is particularly evident in industries where the demand for lightweight, strong, and complex components is paramount.

3.5 INDUSTRY 4.0- ENABLED COMPOSITE MANUFACTURING

Over the past few decades, composites have emerged as the preferred material in the aerospace industry, and there has been a notable increase in their use for aerospace applications in recent years. The American Composite Manufacturers Association's 2020 State of the Industry report highlights the significant growth potential in the carbon fiber composite market, attributed not only to the increased usage of the material but also to the growing number of aircraft launches and deliveries annually. With the rising demand seen in the aerospace sector, there is a pressing need for the industry to embrace digitization and automation of manufacturing processes using cutting-edge technologies. Additionally, the recent shift from nest production to production-line manufacturing underscores the necessity for automated mass production [7, 29].

In a conventional production line, making changes or modifications to a product often necessitates extensive line modifications and manual interventions. Therefore, to make such a product change economically viable, a minimum batch size of the new product is typically necessary. Conversely, in the Industry 4.0 concept, intelligent factories enable the profitable manufacturing of individual customer requirements and even unique items [30]. These factories can readily accommodate last-minute changes on behalf of customers and suppliers. As a result, a key goal of the proposed manufacturing line is

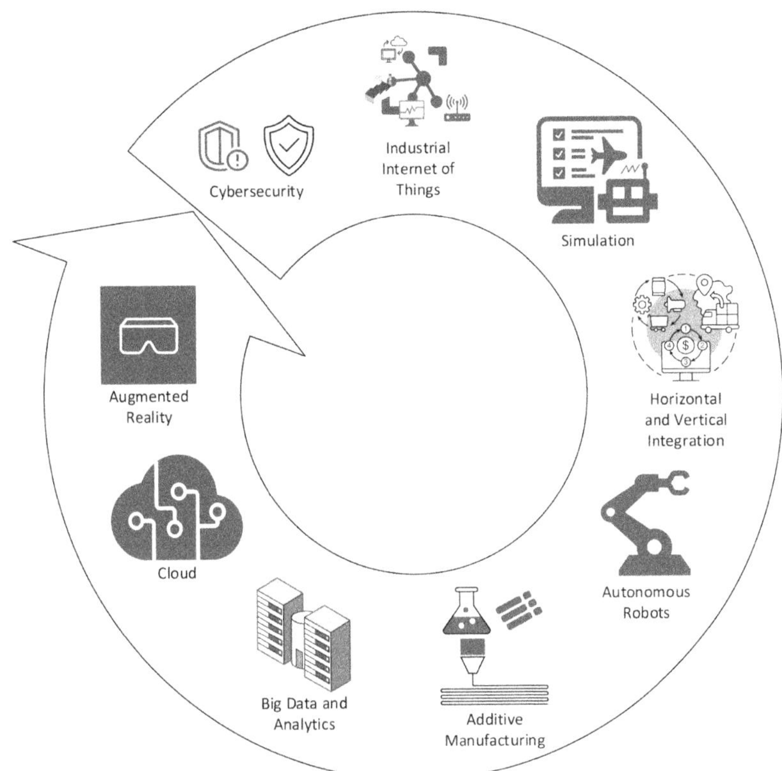

Figure 3.8 Schematic manufacturing line to facilitate the production of various products.

to facilitate the production of various products with minimal or no manual intervention (Figure 3.8). In the context of Industry 4.0's vision for manufacturing, intelligent machines, storage systems, and production facilities interact independently by sharing information, initiating actions, and organizing tasks [31]. As a result, considerable work has been done to facilitate this connectivity among machines at various points along the suggested production line.

3.6 CONCLUSION

In conclusion, the world of advanced manufacturing techniques for composite manufacturing is poised to revolutionize industries globally. Innovative methods such as AM and intelligent material design are reshaping the production of lightweight, durable, and high-performance composites. These techniques not only improve efficiency and precision but also unlock new realms of design and application possibilities. Moreover, beyond immediate benefits, these advanced techniques offer a solution to pressing contemporary challenges. They reduce material waste, energy consumption, and production time, aligning with the global movement toward sustainability. The versatility of composites enables the creation of products that are not just stronger and lighter but also more resilient in demanding environments. Looking forward, the integration of advanced manufacturing with smart technologies like artificial intelligence (AI) and Internet of Things (IoT) will further elevate the efficiency and intelligence of composite manufacturing. This integration promises highly customized, cost-effective solutions, pushing the boundaries of design complexity and performance. The journey toward advanced manufacturing for composites goes beyond perfecting processes; it is about pioneering sustainability, efficiency, and limitless possibilities for future industries. This embodies collaboration, innovation, and ongoing improvement, paving the way for composites to lead technological advancement and industrial progress in the coming era.

3.7 LIMITATIONS AND FUTURE SCOPE

As the field of composite materials continues to evolve, driven by advancements in materials science, manufacturing technologies, and industrial applications, it is imperative to outline the future research directions that will shape the next generation of manufacturing techniques for composites. This chapter presents a roadmap for future research endeavors aimed at enhancing the performance, efficiency, and sustainability of composite manufacturing processes.

- Integrate multiscale modeling techniques to capture the complex interactions between microstructure, process parameters, and final part properties.
- Develop novel recycling and reuse methods for composite waste materials, including thermoset composites, to promote a circular economy in the composites industry.
- Explore the use of 3D printing/AM for fabricating multi-material composite parts with complex geometries and tailored properties.
- Develop joining and bonding methods for assembling hybrid composite structures, ensuring compatibility, durability, and structural integrity.

By addressing the research directions outlined in this chapter, researchers, engineers, and industry stakeholders can unlock new opportunities for innovation and advancement in the field of composite materials, leading to safer, lighter, and more efficient products across various sectors.

REFERENCES

1. D. K. Rajak, P. H. Wagh, and E. Linul, "Manufacturing technologies of carbon/glass fiber-reinforced polymer composites and their properties: A review," *Polymers (Basel)*, vol. 13, no. 21, p. 3721, 2021, doi: 10.3390/polym13213721.
2. J. Frketic, T. Dickens, and S. Ramakrishnan, "Automated manufacturing and processing of fiber-reinforced polymer (FRP) composites: An additive review of contemporary and modern techniques for advanced materials manufacturing," *Additive Manufacturing*, vol. 14, pp. 69–86, 2017, doi: 10.1016/j.addma.2017.01.003.

3. A. Bhardwaj *et al.*, "Additive manufacturing processes for infrastructure construction: A review," *Journal of Manufacturing Science and Engineering*, vol. 141, no. 9, pp. 1–13, 2019, doi: 10.1115/1.4044106.
4. B. Chen, X. Xi, C. Tan, and X. Song, "Recent progress in laser additive manufacturing of aluminum matrix composites," *Current Opinion in Chemical Engineering*, vol. 28, pp. 28–35, 2020, doi: 10.1016/j.coche.2020.01.005.
5. B. Brenken, E. Barocio, A. Favaloro, V. Kunc, and R. B. Pipes, "Fused filament fabrication of fiber-reinforced polymers: A review," *Additive Manufacturing*, vol. 21, no. 2017, pp. 1–16, 2018, doi: 10.1016/j.addma.2018.01.002.
6. S. C. Ligon, R. Liska, J. Stampfl, M. Gurr, and R. Mülhaupt, "Polymers for 3D printing and customized additive manufacturing," *Chemical Reviews*, vol. 117, no. 15, pp. 10212–10290, 2017, doi: 10.1021/acs.chemrev.7b00074.
7. H. Parmar, T. Khan, F. Tucci, R. Umer, and P. Carlone, "Advanced robotics and additive manufacturing of composites: Towards a new era in Industry 4.0," *Materials and Manufacturing Processes*, vol. 37, no. 5, pp. 483–517, 2022, doi: 10.1080/10426914.2020.1866195.
8. G. D. Goh, Y. L. Yap, S. Agarwala, and W. Y. Yeong, "Recent progress in additive manufacturing of fiber reinforced polymer composite," *Advanced Materials Technology*, vol. 4, no. 1, pp. 1–22, 2019, doi: 10.1002/admt.201800271.
9. H. Ikram, A. Al Rashid, and M. Koç, "Additive manufacturing of smart polymeric composites: Literature review and future perspectives," *Polymer Composites*, vol. 43, no. 9, pp. 6355–6380, 2022, doi: 10.1002/pc.26948.
10. X. Wang, M. Jiang, Z. Zhou, J. Gou, and D. Hui, "3D printing of polymer matrix composites: A review and prospective," *Composites: Part B, Engineering*, vol. 110, pp. 442–458, 2017, doi: 10.1016/j.compositesb.2016.11.034.
11. P. Zhuo, S. Li, I. A. Ashcroft, and A. I. Jones, "Material extrusion additive manufacturing of continuous fibre reinforced polymer matrix composites: A review and outlook," *Composites: Part B, Engineering*, vol. 224, no. January, p. 109143, 2021, doi: 10.1016/j.compositesb.2021.109143.
12. Z. Liu, Y. Wang, B. Wu, C. Cui, Y. Guo, and C. Yan, "A critical review of fused deposition modeling 3D printing technology in manufacturing polylactic acid parts," *The International Journal, Advanced Manufacturing Technology*, vol. 102, no. 9–12, pp. 2877–2889, 2019, doi: 10.1007/s00170-019-03332-x.
13. C. Liu *et al.*, "3D printing technologies for flexible tactile sensors toward wearable electronics and electronic skin," *Polymers (Basel)*, vol. 10, no. 6, p. 629, 2018, doi: 10.3390/polym10060629.
14. X. Tian, A. Todoroki, T. Liu, L. Wu, Z. Hou, M. Ueda, Y. Hirano, R. Matsuzaki, K. Mizukami, K. Iizuka, A. V. Malakhov, A. N. Polilov, D. Li, and B. Lu, 3D printing of continuous fiber reinforced polymer composites: Development, *Application, and Prospective Chinese Journal of Mechanical Engineering: Additive Manufacturing Frontiers*, 1 (2022) 100016, https://doi.org/10.1016/j.cjmeam.2022.100016.
15. S. W. Paek, S. Balasubramanian, and D. Stupples, "Composites additive manufacturing for space applications: A review," *Materials (Basel)*, vol. 15, no. 13, p. 4709, 2022, doi: 10.3390/ma15134709.
16. D. Vaes and P. Van Puyvelde, "Semi-crystalline feedstock for filament-based 3D printing of polymers," *Progress in Polymer Science*, vol. 118, p. 101411, 2021. https://doi.org/10.1016/j.progpolymsci.2021.101411.
17. V. Monfared, H. R. Bakhsheshi-Rad, S. Ramakrishna, M. Razzaghi, and F. Berto, "A brief review on additive manufacturing of polymeric composites and nanocomposites," *Micromachines*, vol. 12, no. 6, p. 704, 2021, doi: 10.3390/mi12060704.
18. A. Dickson, H. Abourayana, and D. Dowling, "Composites using fused filament fabrication: A review," *Polymers (Basel)*, vol. 12, no. 2188, pp. 1–18, 2020.

19. S. J. Kalita, S. Bose, H. L. Hosick, and A. Bandyopadhyay, "Development of controlled porosity polymer–ceramic composite scaffolds via fused deposition modeling," *Materials Science & Engineering. C*, vol. 23, no. 5, pp. 611–620, 2003, doi: 10.1016/S0928-4931(03)00052-3.

20. O. S. Carneiro, A. F. Silva, and R. Gomes, "Fused deposition modeling with polypropylene," *Materials & Design*, vol. 83, pp. 768–776, 2015, doi: 10.1016/j.matdes.2015.06.053.

21. C. Bellehumeur, L. Li, Q. Sun, and P. Gu, "Modeling of bond formation between polymer filaments in the fused deposition modeling process," *Journal of Manufacturing Processes*, vol. 6, no. 2, pp. 170–178, 2004, doi: 10.1016/S1526-6125(04)70071-7.

22. A. N. Dickson, H. M. Abourayana, and D. P. Dowling, "3D printing of fibre-reinforced thermoplastic composites using fused filament fabrication: A review," *Polymers*, 2020, vol. 12, 2188, doi:10.3390/polym12102188.

23. W. Liu, Z. Sun, H. Ren, X. Wen, W. Wang, T. Zhang, L. Xiao, and G. Zhang. "Research progress of self-healing polymer for ultraviolet-curing three-dimensional printing," *Polymers*, 2023, vol. 15, 4646. https://doi.org/10.3390/polym15244646.

24. J. P. Kelly, J. W. Elmer, F. J. Ryerson, J. R. I. Lee, and J. J. Haslam, "Directed energy deposition additive manufacturing of functionally graded Al-W composites," *Additives Manufacturing*, vol. 39, p. 101845, 2021, doi: 10.1016/j.addma.2021.101845.

25. D. Wu *et al.*, "One-step additive manufacturing of TiCp reinforced Al_2O_3–ZrO_2 eutectic ceramics composites by laser directed energy deposition," *Ceramics International*, vol. 49, no. 8, pp. 12758–12771, 2023, doi: 10.1016/j.ceramint.2022.12.141.

26. L. Chen, R. Wilson, G. de Looze, K. Yang, and A. Seeber, "Additive manufacturing of titanium alloy based composites using directed energy deposition: Study of microstructure, mechanical properties and thermal conductivity," *Materials Science Engineering A*, vol. 884, no. June, p. 145579, 2023, doi: 10.1016/j.msea.2023.145579.

27. A. Dass and A. Moridi, "State of the art in directed energy deposition: From additive manufacturing to materials design," *Coatings*, vol. 9, no. 7, p. 418, 2019, doi: 10.3390/coatings9070418.

28. K. T. Cho, L. Nunez, J. Shelton, and F. Sciammarella, "Investigation of effect of processing parameters for direct energy deposition additive manufacturing technologies," *Journal of Manufacturing and Materials Processing*, vol. 7, no. 3, p. 105, 2023, doi: 10.3390/jmmp7030105.

29. M. Stojkovic and J. Butt, "Industry 4.0 implementation framework for the composite manufacturing industry," *Journal of Composites Science*, vol. 6, no. 9, p. 258, 2022, doi: 10.3390/jcs6090258.

30. R. Ashima, A. Haleem, S. Bahl, M. Javaid, S. Kumar Mahla, and S. Singh, "Automation and manufacturing of smart materials in additive manufacturing technologies using Internet of Things towards the adoption of Industry 4.0," *Materials Today. Proceedings*, vol. 45, pp. 5081–5088, 2021, doi: 10.1016/j.matpr.2021.01.583.

31. M. Ryalat, H. ElMoaqet, and M. AlFaouri, "Design of a smart factory based on cyber-physical systems and Internet of Things towards Industry 4.0," *Applied Sciences*, vol. 13, no. 4, p. 2156, 2023, doi: 10.3390/app13042156.

Chapter 4

Evaluation of mechanical properties of Mg/CNT/Al₂O₃-based metal matrix nanocomposites using stir casting process

Bhaskar Chandra Kandpal, Gaurav Sharma, Nitin Johri, Natraj Mishra, and Ajay Kumar

4.1 INTRODUCTION

Bio-implants are prostheses used to normalize physiological functions. Biosynthetic materials, such as collagen, and tissue-engineered goods, such as artificial skin or tissues, are used to create them. In comparison to conventional bio-implant materials, magnesium and its alloys are now extensively employed in bio-implants due to their biodegradability and high mechanical qualities. Bio-implants are made using a variety of manufacturing technologies, including powder metallurgy, friction stir processing, additive manufacturing, stir casting, and so on. Presently, stir casting is the most widely used commercial and cost-effective method for producing magnesium-based metal matrix composites (MMCs). Mechanical stirring is frequently used to incorporate powder reinforcement into molten metal during this technique. The utilization of a wide spectrum of reinforcement and matrices allows composites to have a lot of design freedom. As a consequence, we can get hybrid properties in a single material, which is impossible to get with conventional materials. Magnesium-based matrices are high-strength metals that are still being used today. However, pure magnesium is rarely used in structural and biomedical applications because of poor mechanical properties. The addition of aluminum and zinc to magnesium improves its strength and castability. The addition of manganese helps in conception. These alloys are strong, rigid, and corrosion-resistant at low and high temperatures. The reinforcement should be stronger and stiffer than the matrix to get the desired strengthening effect. When compared with aluminum matrix composites, the research for magnesium matrix composites is very limited.

Magnesium matrix composites reinforced with carbon nanotube (CNT) potential in transportation are studied in Reference [1] with the analysis that its application faces the main obstacle on account of low absolute strength. Composites were prepared using AZ61 powder and copper-coated CNTs as raw materials, having superior interfacial bonding, fine-grained and uniformly distributed carbon nanotubes through refinement by ball milling process of AZ61 powder, hot press sintering, and dispersion by ball milling. CNTs are seen to have uniform dispersal while their volume fraction is limited to 1%. However, with higher content of CNTs, compressive and yield strength increases. The yield strength and compressive strength of these composites reach up to 361 MPa and 439 MPa, respectively, with a 1% volume fraction of CNTs. These levels are seen to be 9% and 14% higher with similar fracture strain for composites with less than 1% volume fraction of CNTs. The clustering of CNTs results in reduced strength, and fracture strain is seen when the volume fraction exceeds 1%. Magnesium-based MMCs are commonly used in bio-implants and other biomedical applications [2]. Synthesis of these composites is done by various methods of processing such as additive manufacturing, stir casting, powder metallurgy, friction stir processing, and so on. An economical method of fabrication using stir casting method is preferred on account of simplicity, mass production capacity, low production cost, and established process. Important aspects of the stir casting process include the design of the furnace, selection of material, and quality control which are looked at in detail. Composite manufacturing-related research prospects, characteristics of composites, and manufacturing obstacles were also focused. Advice on reinforcement material selection, design of the furnace, procedure parameters, and additives was given resulting in a unique evaluation. A summary of commonly used reinforcements for bio-implants and technological advancements related to stir casting are discussed to aid researchers in magnesium matrix composite-based bone implants and fabrication methods. Magnesium provides improved strength and stiffness on account of being light and capacity to replace alloys used in applications that require lightweight along with increased strength and stiffness [3]. Inclusion of reinforcing material in metallic matrix provides a significant impact on wear behavior, stiffness, creep resistance, specific strength, and damping behavior in comparison to general-purpose engineering materials. These superior mechanical and physical characteristics of magnesium matrix composites make them viable for different applications. Appropriate methods and respective process parameters for composite synthesis are studied overviewing reinforcement impact in magnesium and alloys, emphasizing advantages and limitations. Critical examination of magnesium-based MMCs' essential applications and reinforcement impact on mechanical characteristics and microstructure were also focused on. Composites of aluminum alloy 7148 with CNT (multiwalled) in combination with liquefied metals enable intricate design creation along with fabrication for mass production using heat treatment, mechanical stir casting, and thermoforming [4]. Optimum parameter investigation was done by choosing two levels of factorials in Taguchi and affect variables like wettability of magnesium, the concentration of nanotube, and stir duration of ring mechanical, in addition to the superior design of experimentation. Parameters like hardness, S/N ratio, and tensile strength (ultimate) were selected as response variables. The design of the experimentation run resulted in the fabrication of a nanocomposite with optimized features having improved tensile strength and hardness. Demonstrations were done combining heat treatment with thermoforming resulting in mechanical property improvement for the fabrication of multiwalled CNTs using stir casting. A review of aluminum MMCs' mechanical properties and surface characteristics like flexural and tensile strength, hardness, impact strength, and wear resistance was done with the findings that the processing in liquid state was more popular than solid state on account of reinforcement particles better dispersion in matrix material [5]. Stir casting finds common application as a processing method in a liquid state on account of low production cost and ease. Surface and mechanical characteristics are seen to improve by various reinforcement particle additions in small percentages. The working parameter's influence on aluminum MMCs' wear rate was also analyzed. Variation of wear loss along with impact resistance is seen to be crucial for automotive and aerospace applications. Novel fabrication method strategies of magnesium matrix-based composites with reinforcements of CNTs, nanoceramic particles, and nanoplatelets of graphene, composite property characterization, corrosion behavior, mechanical response, and applications were discussed [6], with an aim toward the investigation of the relevant studies along with a conclusion. Complex challenge-handling strategies are suggested for this class of composites' potential usage as structural material functional requirements in the automotive and aerospace industries. The challenges come in the form of reinforcement like CNTs' uniform distribution in a matrix of magnesium which is dependent on fabrication process parameters. Recycling ease, lightweight, and superior specific strength along with high thermal conductivity, surface-to-volume ratio, tensile strength, and low density give considerable focus to the application of these composites in automotive, medical, and aerospace industries. A powder metallurgy-based magnesium matrix reinforced with CNTs (multiwalled) was prepared [7]. The ball milling process was followed for mixing of powders with the rotational speed of 60 rpm for a 6-hour duration with sintering for a 2-hour duration at 530°C. The volume percentage of CNTs in magnesium ranged from 10%, 20%, 40%, and 60%. Measurement of mechanical and physical properties like compressive strength, density, hardness, porosity, and wear rate was done along with comparison with sample values of pure magnesium (unmodified). Lightweight composite manufacturing ability was referred to by density results. Mechanical tests of CNTs in magnesium matrix composites reveal significant improvement in compressive strength. Optimized compressive strength at 0.4% volume fraction of CNT is seen. Microhardness values do not seem to have much effect due to variations in the percentage of CNT. An increase in wear resistance with a volume fraction of CNT reinforcement is also seen. Wear characteristics were analyzed for stir-casting-based magnesium matrix composites [8]. Pin on disk machine was used under dry sliding for the conduct of wear test, with the counter face of hardened steel, along with sliding speed variation ranging from 200 rpm, 400 rpm, 600 rpm, and 800 rpm and load application of 10/20 N at room temperature with 0–180 seconds time interval. The addition of nanoparticles of alumina resulted in a decrease in wear at increased sliding speed under normal load along with a substantial rise in coefficient of friction value increase to 0.55 from 0.095. Mechanical stir casting effect on the lightweight alloy of aluminum (7075 based) was analyzed along with the effect of 20–30 nm average-sized nanoparticles of Al_2O_3 with 10%, 20%, 30%, and 40% weight fractions [9]. Nanoparticles' non-consistent distribution is seen to lead toward a highly porous matrix. The electric version of stir casting is suggested for protection against this problem encountered in the mechanical version with an infusion of particulates of Al_2O_3 directly in aluminum alloy with a 1% weight fraction. Reinforcement wettability is improved by induction of powder of micro-magnesium with a 1% weight fraction. Composites evaluation was subsequently carried out by SEM, optical microscopy, and so on. Consistent distribution of nanoparticles throughout the matrix was confirmed by micrographs of SEM. along with performance analysis of the microstructure of active grain. Compared to a base alloy of Al7075, the impact and tensile strengths, as well as hardness are improved with 1% weight fraction of nano Al_2O_3 particles-reinforced alloy. Processing techniques were analyzed for aluminum matrix composites reinforced with CNTs finding wide applications in automotive and aerospace industries, on account of production challenges encountered [10]. Mechanical properties of this class of composites are enhanced based on the technique of processing employed with a review of solid-state processing for synthesis of composites, for insight into techniques capabilities and features with regard to CNT reinforcement incorporation. In summary, the aluminum–CNT composites mechanical performance is based on the technique's capability in CNT dispersion within the matrix of aluminum. Good formability coupled with other superior properties like lightness, heat and electrical conductivity, low toxicity, and corrosion resistance are seen to be the contributing properties for the widespread application of this class of composites. High stiffness, with specific strength, is required for their application in structural applications along with aluminum-strengthening requirements.

The literature reviewed shows the growing acceptance of magnesium matrix-based composites on account of their high specific strength, low density, recycling ease, and other favorable characteristics. Research in this domain will help in handling challenges faced in fabrication, damage through corrosion, wear, and so on, and process parameter optimization.

- Mg and its alloys have been identified as promising bone implant materials owing to their natural degradability, good biocompatibility, and favorable mechanical properties.
- Nevertheless, the too fast degradation rate usually results in a premature disintegration of mechanical integrity and local hydrogen accumulation, which limit their clinical bone repair application.
- So magnesium-based MMCs will be developed to overcome this limitation of Mg and its alloys for bone implants.

4.2 EXPERIMENTAL METHODOLOGY

4.2.1 Preparation of samples

For the manufacturing of magnesium composites, mechanical stir casting is utilized, which is further improved by adding reinforcement, such as CNTs and aluminum oxide (Al_2O_3), to the molten magnesium alloy. In comparison to other methods for fabricating MMCs, this approach is both cost-effective and straightforward. Figure 4.1 depicts the stir-casting setup utilized in this project. Here the material was melted using an electric furnace at the required temperature of over 550°C.

Magnesium alloy ingots were chopped into little pieces, each weighing about 1 kg, and then melted in a crucible with the assistance of an electric furnace. Pure magnesium alloy pieces were melted in a crucible with the assistance of an electric furnace at a temperature of roughly 550–650°C for 2 hours to achieve a liquid condition. Flux (approximately 10% by weight) was introduced before the stirring phase to prevent the magnesium from exploding. To prevent magnesium from oxidizing, argon gas was supplied throughout the melting process. The reinforced materials were stirred in with the molten magnesium for 10–15 minutes at 650°C. The stirring procedure was carried out with the assistance of a rotor that rotated the stirrer at a rate of 200–250 rpm. The melting temperature in the furnace was around 700°C throughout this operation. The slurry was then poured into the die to achieve the required form, as shown in Figure 4.2.

Figure 4.1 Mechanical stir casting process: stir casting setup for manufacturing magnesium composites.

Figure 4.2 Magnesium alloy sample.

Additional samples containing various reinforcement compositions, including aluminum oxide (Al_2O_3) and CNTs, underwent the same processing procedure. Figure 4.3 shows the melting process of magnesium alloy in electric furnace. With the help of this furnace, additional material can be melted.

4.2.2 Quality of flux to prevent Mg from oxidation

The weight percentages of chemicals required to create flux is about 10% of the total weight of the sample. Table 4.1 shows the composition of flux for

Figure 4.3 Molten metal: Melting process of magnesium alloy in electric furnace.

Table 4.1 Composition of flux for magnesium casting

Sl. No.	Chemical	Percentage (%)
1	MgO	15
2	KCl	20
3	CaF_2	15
4	$MgCl_2$	50

Table 4.2 Chemical compositions of Mg-CNT and Mg-CNT-A_2O_3 matrix

Sl. No.	Composition	Weight (grams)	CNT reinforcement (grams)	Al_2O_3 reinforcement (grams)
1	Pure Mg	390	Nil	Nil
2	Pure Mg + CNT (0.5%)	714	4.04	Nil
3	Pure Mg + CNT (0.3%) + Al_2O_3 (2%)	670	2.45	13.52

magnesium casting. Table 4.2 depicts the chemical compositions of Mg-CNT and Mg-CNT-A_2O_3 matrix.

4.2.3 Tensile test

The cast MMC plate was cut using an EDM machine following lathe machining. The ASTM E8 subsize specification was followed in the manufacturing of the tensile test specimen on an EDM machine. Each plate was used to make two tensile test specimens; sample 2 was used if sample 1 failed due to an internal imperfection in order to provide a more accurate result. Figure 4.4 displays specimens ready for tensile testing of MMCs based on magnesium.

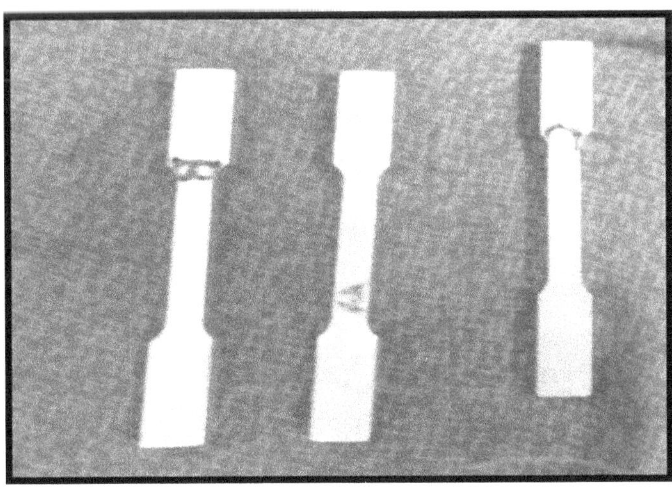

Figure 4.4 Tensile test specimen. Specimen preparation for tensile testing of magnesium-based metal matrix composites.

4.3 RESULTS AND DISCUSSION

4.3.1 Mechanical testing

To determine the mechanical properties of vacuum-assisted stir die-cast composites, a tensile test of MMCs augmented with SiC particles was conducted. Tensile samples were made using commercial magnesium alloy AZ91 and composites containing varying proportions of SiC particles. As per the ASTM E8/E8M standard, the tensile sample's gauge length was 20 mm, and its gauge diameter was 8 mm. Tensile tests were performed using the Instron-4208 universal testing equipment at room temperature and a starting strain rate of 0.005 seconds.

The stress–strain curves for the reinforced magnesium alloys and their composite are shown in Figure 4.5. Figure 4.5a presents the stress–strain curve representing the behavior of a magnesium alloy without any reinforcement. Figure 4.5b depicts the stress–strain behavior of a magnesium alloy reinforced with CNTs. This curve demonstrates how the addition of CNT reinforcement influences the material's mechanical response. In Figure 4.5c, the stress–strain curve represents the behavior of a magnesium alloy reinforced with both CNTs and aluminum oxide (Al_2O_3) particles. This composite exhibits a combined effect of the two types of reinforcement.

4.3.2 Hardness test

To determine the homogeneity of the reinforcement particle distribution and the variance in hardness with varying reinforcement percentages, a hardness test was conducted on each composite. A portion of polished composites was subjected to the Vickers micro-hardness test with a 1.0 kg load with a dwell time of 10 seconds. The macro-hardness test has also been performed on the composite using a load of 3 kg with a dwell time of 10 seconds. LECO's LV Series Macro-Vickers Hardness machine is used for testing hardness. For the manufacturing of magnesium composites, mechanical stir casting is utilized, which involves adding reinforcement, Table 4.3 shows the results

Table 4.3 Comparison of tensile strength for Mg-MMCs

Sl. No.	Composition	Reinforcement percentage	Tensile strength (kg/mm^2)
1	Pure Mg	Nil	3.426
2	Pure Mg + CNT (0.5%)	0.5	6.887
3	Pure Mg + CNT (0.3%) + Al_2O_3 (2%)	2.3	7.047

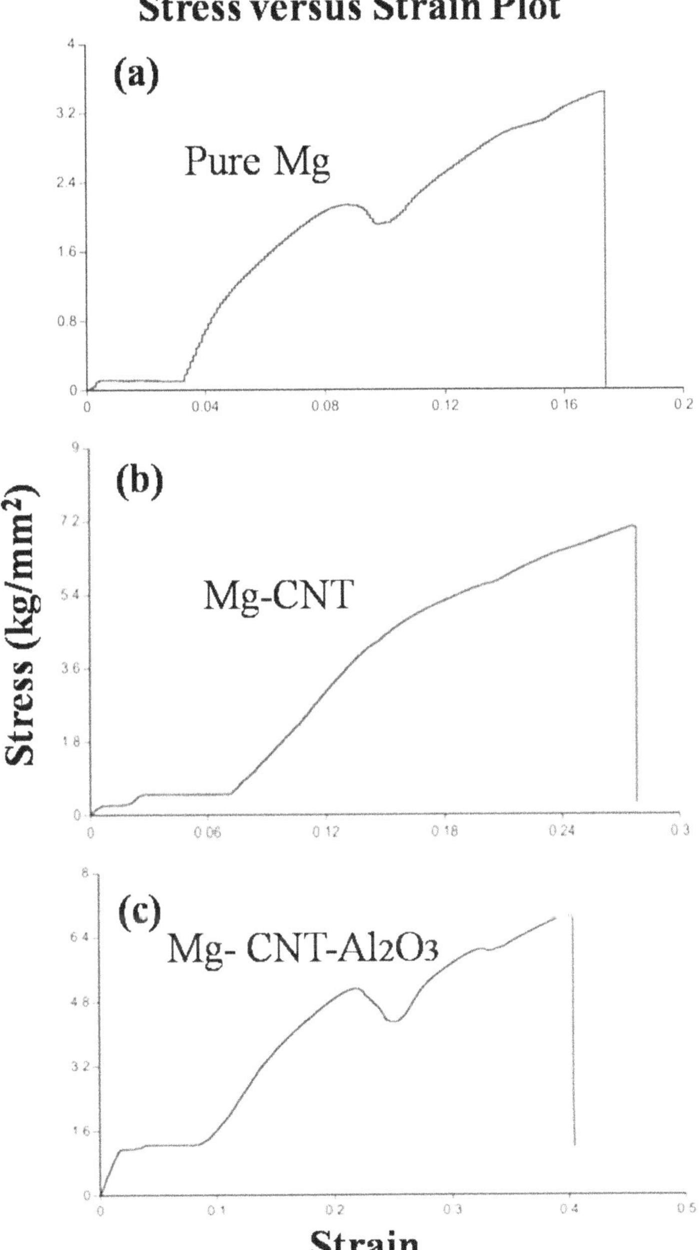

Figure 4.5 Stress–strain curves for the reinforced magnesium alloys and their composite. (a) Pure Mg alloy composite. (b) Mg-CNT alloy composite. (c) Mg-CNT-Al$_2$O$_3$ alloy composite.

Table 4.4 Hardness comparison of magnesium-based metal matrix composites with different reinforcement percentages

Composition	Reinforcement percentage	Load applied (kg)	Hardness value (kg/mm²)
Pure magnesium	Nil	10	82.2
Magnesium + CNT	0.5	10	85.4
Pure magnesium + CNT + Al₂O₃	2.3	10	97.1

of hardness testing on each specimen after the casted samples have been finished. Vickers hardness was determined in this study, that is, the Vickers scale was used to characterize the indenter penetration scale, which was loaded into the visible test pieces. Each sample has a 10-kg applied load and a four-digit digital scale.

The results shown in Table 4.4 represent an improvement in hardness when CNT is added to molten magnesium during the casting process.

4.3.3 Impact strength

The impact resistance strength of three samples is shown in Tables 4.5 and 4.6, which is an ASTM standard technique for evaluating the impact resistance of materials. The starting scale is 164 joules.

Table 4.5 Comparison of Impact strength (IZOD-Test) for Mg-MMC

Composition	Izod value (joules)
Pure magnesium	18
Magnesium + CNT (0.5%)	164
Magnesium + CNT (0.3%) + Al₂O₃ (2%)	20

Table 4.6 Comparison of impact strength (CHARPY-Test) for Mg-MMC

Composition	Charpy value (joules)
Pure magnesium	4
Magnesium + CNT (0.5%)	10
Magnesium + CNT (0.3%)+Al₂O₃ (2%)	2

4.4 RESULTS AND DISCUSSION

The casted samples of magnesium-based MMCs reinforced with CNTs and aluminum oxide (Al_2O_3) exhibited notable improvements in tensile strength and hardness compared to the base alloy. This enhancement can be attributed to the reduction of MgO and residual strain in the composite, leading to improved interfacial strengthening, particularly with the formation of Al_2MgC_2 ternary carbide. These findings were consistent with the observed improvements in impact resistance. The experimental results align with the broader conclusions drawn from the study, indicating that tensile strength and hardness increase with higher weight percentages of reinforcement, while ductility and elongation decrease. The presence of modest amounts of magnesium resulted in porosities within the MMCs, highlighting a challenge for further optimization. Additionally, inadequate bonding between CNT/Al_2O_3 particles and the magnesium matrix was observed, contributing to discontinuous chip creation and increased brittleness with higher reinforcement percentages. Furthermore, SiC particle-reinforced magnesium MMCs demonstrated superior wear and creep resistance compared to Al_2O_3-reinforced MMCs, suggesting the potential for tailored material properties through reinforcement selection. Overall, the incorporation of CNTs into the magnesium matrix improved wettability and bonding strength, enhancing sliding wear resistance and positioning magnesium MMCs as promising materials for various industrial applications. There was an improvement in tensile strength and hardness of casted samples of Mg-based MMC with CNT and aluminum oxide as reinforcements. It was due to the elimination of MgO and less residual strain in the composite. The improvement in mechanical properties was due to the interfacial strengthening of Al_2MgC_2 ternary carbide. Similar findings were improved in the impact test also.

4.5 CONCLUSION

This study confirms the potential of magnesium-based MMCs for various applications, particularly in sectors such as automotive, aerospace, and biomedical engineering. The experimental findings demonstrate that the mechanical properties of magnesium alloys can be significantly enhanced through the incorporation of suitable reinforcements like CNTs and Al_2O_3 particles. However, challenges such as inadequate bonding between reinforcement particles and the magnesium matrix were observed, suggesting areas for further improvement in fabrication processes. Despite these challenges, the developed composites exhibit promising characteristics, including enhanced tensile strength, hardness, and wear resistance, which could lead to their widespread adoption in lightweight structural and functional components across different industries.

4.6 LIMITATIONS

To address global concerns about the scarcity of environmentally acceptable materials, extensive research in the fields of materials and metallurgical science is required. The need to create recyclable nanocomposites based on magnesium has increased due to the metal's susceptibility to corrosion and flammable materials. The oxidation of pure magnesium has been addressed by a variety of techniques, such as the creation of composites with suitable organic reinforcing agents, metal alloys, or surface modifications. Magnesium is the lightest metal ever found and has great potential use in structural alloys. Due to their lightweight nature, enhanced performance, reduced weight, improved rigidity, toughness, strength-to-weight proportion, and fatigue resistance, magnesium-based MMCs are the best choices and have so far resolved all mechanical and biomedical application challenges. The automotive and aerospace industries are drawn to the lighter alloys and nanocomposites, which provide an alternative to heavy metals like steel, cast iron, zinc compositions, and even aluminum alloys. These days magnesium-based composites are also developed for biomedical applications using the additive manufacturing technique.

REFERENCES

1. Ding Y, Shi Z, Li Z, Jiao S, Hu J, Wang X, Zhang Y, Wang H, and Guo X. 2022. Effect of CNT content on microstructure and properties of CNTs/refined-AZ61 magnesium matrix composites. *Nanomaterials* **12**, 2432.
2. Chandra Kandpal B, Johri N, Kumar L, Tyagi A, Joshi V, and Gupta U. 2022. Stir casting technology for magnesium-based metal matrix composites for bio-implants: A review. *Mater. Today Proc.* **62**, 4519–25.
3. Arora G S, Saxena K, Mohammed K A, Prakash C, and Dixit S. 2022. Manufacturing techniques for Mg-based metal matrix composite with different reinforcements. *Crystals* **12**, 945.
4. Refaai M R A, Ravi S, Prasath S, Thirupathy M, Subbiah R, and Diriba A. 2022. Optimization of stir casting variables for production of multiwalled carbon nanotubes: AA7149 composite. *J. Nanomater.* **2022**. https://doi.org/10.1155/2022/2535470
5. Kumar D, Angra S, and Singh S. 2022. Mechanical properties and wear behaviour of stir cast aluminum metal matrix composite: A review. *Int. J. Eng. Trans. A: Basics* **35**, 794–801.
6. Abazari S, Shamsipur A, Bakhsheshi-Rad H R, Ismail A F, Sharif S, Razzaghi M, Ramakrishna S, and Berto F. 2020 Carbon nanotubes (CNTs)-reinforced magnesium-based matrix composites: A comprehensive review. *Materials (Basel)* **13**, 1–38.
7. Abed Zaid H M, Abed A R N, and Hasan H S. 2019. Physical and mechanical properties of magnesium (Mg) matrix composites reinforced with carbon nanotube (CNT) fabricated by powder metallurgy technique. *J. Eng. Appl. Sci.* **14**, 10087–91.
8. Thirugnanasambandham T, Chandradass J, Baskara Sethupathi P, and Leenus Jesu Martin M. 2019. Experimental study of wear characteristics of Al_2O_3 reinforced magnesium based metal matrix composites. *Mater. Today Proc.* **14**, 211–8.

9. Suresh S, Gowd G H, and Deva Kumar M L S. 2019. Experimental investigation on mechanical properties of Al 7075/Al$_2$O$_3$/Mg NMMC's by stir casting method. *Sādhanā* **44**, 21.
10. Toozandehjani M, Matori K A, Ostovan F, Abdul Aziz S, Mamat M S, and Oskoueian A. 2017. Carbon nanotubes reinforced aluminum matrix composites: A review of processing techniques. *Pertanika J. Sch. Res. Rev.* **3**, 7092.

Chapter 5

Utilization of industrial waste as filler material in the development of polymer composites

*Alok Agrawal, Abhilash Purohit, Gaurav Gupta,
Sandip Kumar Nayak, Pravat Ranjan Pati,
and Alok Satapathy*

5.1 INTRODUCTION

For the last many years, the quantity and types of waste have been increasing due to the enhancement of urbanization and growth in the economy because of the implementation of innovative technology in the industries. The waste is generated by various industries, including waste from mining industries, agricultural activities, and domestic waste. Out of the total waste generated from different sectors, around 90% of the waste is from industries. Among the various waste generated by different industries, the maximum amount of waste is disposed of in nearby land or water bodies and because of that the environmental balance is disturbed. Disposing of the solid hazardous waste properly so that the negative impact can be minimized has become a worldwide serious issue. It is a massive challenge before the research community to utilize the mentioned waste to ensure sustainable development keeping in mind the cost concern of the present material [1]. It is now compulsory for all industries to perform chemical treatment of the waste before disposing it so that the harmful effect of such waste can be nullified. It is to be noted that the cost involved in the chemical treatment of the waste is large because of the complicated process involved in it. Apart from that, the by-product of the chemical treatment releases harmful gases and chemicals, which are again harmful to the environment; hence, treatment of waste before disposal is not a proper solution. Furthermore, the area needed to hold the waste will remain unchanged even after chemical treatment. Hence, recycling or reuse of such waste is considered to be the most appropriate solution to get rid of such a problem [2].

Blast furnace slag (BFS), LD slag, and LD sludge are steel industry waste. BFS is generated during the process of abstraction of iron from its raw material, which is called ore in the blast furnace (BF). LD slag is generated by the LD converter. LD sludge is also a by-product of LD converter and it is present in the lump form. Copper slag is a by-product obtained from copper-making industry. Red mud is the by-product of Bayer's process which is used to obtain alumina. All this waste is generated in huge quantities and is of real concern as their impact on the environment is huge.

DOI: 10.1201/9781003564355-5

For the last many years, all this waste generated from different industries has been used by civil engineers for making roads. This waste is also mixed with Portland cement for the development of concrete. Further, it is used for preparing abrasive tools and ballast, colouring agent for paints, and many others. Despite the increasing usage of all this industrial waste in such large quantity, a still huge amount of waste is getting piled up as their annual production is increasing year by year. Because of this, scientists and material designers all around the world have focused on the development of a new class of polymeric composite material by taking these industrial wastes as reinforcement phases. Waste generated by diverse industries falls under the classification of inorganic fillers; it seamlessly amalgamates with polymers to fashion composite structures. These fillers play a pivotal role in augmenting the mechanical, thermal, and tribological attributes of polymers [1]. Extensive research within the scientific community has focused on integrating such waste as fillers in polymeric resins. This chapter encapsulates previous endeavours, delineating the utilization of industrial waste as filler material for polymer composite development. By providing a benchmark and foundational framework, the chapter facilitates the scientific community in conducting further research within the same domain. This collective effort aims to propel the potential applications of industrial waste-infused polymeric composites across diverse sectors.

5.2 POLYMER MATRIX COMPOSITES FILLED WITH INDUSTRIAL WASTE

5.2.1 Utilization of BFS as filler in polymeric matrix

A non-metallic by-product of making pig iron in a BF is termed BFS. It is created during the pig iron manufacturing process using limestone, iron ore, and coke combustion waste, among other elements. In the process of manufacturing iron, coke is used as fuel in a BF together with iron ore, iron waste, and fluxes (limestone and dolomite). The iron ore is converted to a molten iron product by the combustion of the coke, which produces carbon monoxide. At the bottom, the pig iron is visible above the molten slag. Its temperature, which ranges from 1,400°C to 1,600°C, is comparable to that of the hot, molten iron. Occasionally, the slag rises to the surface and is tapped off [2]. According to chemical analysis, BFS may contain oxides of calcium, silicon, magnesium, aluminium, iron, and sulphur [3].

By combining BFS, glass fibre, with epoxy (EP) resin, Padhi and Satapathy [4] developed composite materials (CMs). Various specimens were produced using a basic hand lay-up technique, each containing different BFS contents (0–30 wt.%). The micro-hardness of the epoxy–glass fibre composite significantly increased in correlation with the quantity of BFS particles in the mixture. The BFS content improved from 0% to 20%, leading to a rise in micro-hardness from 54.37 MPa to 82.67 MPa. Additionally, samples were exposed to solid particle erosion utilizing an erosion testing apparatus resembling an air jet. Erosion tests were conducted utilizing a meticulously developed experimental program that followed the Taguchi technique. Variables such as BFS wt.%, erodent temperature, impingement angle, and impact velocity had a notable effect on reducing erosion rates (E_r). The study showed that the erodent size had the least impact on the E_r of these CMs. The study showed that BFS enhances the composite's resistance to wear and erosion because of its effective filler properties. The study utilized artificial neural network (ANN) to precisely analyse and predict the wear behaviour of the CMs in different experimental conditions, including those beyond the range of study.

Tribological applications for short glass fibre (SGF)-filled polymer CMs are frequent. By adding micro-sized BFS particles, Padhi and Satapathy [5] aimed to increase the wear resistance of EP base CMs filled with SGF. The short glass fibres (GF) utilized in this work have an aspect ratio of 500, a length of roughly 10 mm, and a radius of 0.01 mm. Using a simple hand lay-up method, two sets of CMs were developed with and without 20 wt.% SGF reinforcement in EP resin. The samples were made with varying wt.% of BFS (0, 10, 20, and 30 wt.%). The effects of different input variables on the sliding wear (S_r) property of these CMs were thoroughly investigated. The Taguchi orthogonal arrays (L_{16}) design of the experiment method was applied. Specific wear rates were recorded for each pair of composites throughout all 16 test cycles, along with the appropriate S/N ratios. The total mean S/N ratios for EP/GFCMs were found to be −1.4470 and −6.3930 dB, respectively. This was achieved with the use of the experiment design application-specific software Minitab 14. Based on the S/N ratio of response analysis, sliding velocity was found to be the major variable, followed by GF wt.%. They identified the optimal parameter settings that reduced S_r by utilizing the Taguchi approach. This research further investigates the potential of employing ANNs to predict the sliding wear features of CMs. For EP/BFS/SGF CMs, the % errors between the ANN result and the test values were very small, ranging from 0% to 6%. This suggests a strong agreement between the ANN results and the experimental data. The neural network demonstrated strong generalization capabilities in predicting wear rate based on various input data, different from the original training dataset. This was supported by simulated findings illustrating the influence of key process variables on wear rate. Subsequently, SEM was utilized to examine the morphology of worn surfaces and investigate probable wear mechanisms such as wear tracks, plastic deformation, surface cracks, fracture of short fibres, and BFS particle detachment features.

Additionally, they examined and documented the wear response of a novel class of polypropylene (PP)/SGF/BFS hybrid CMs [6]. In this study, injection moulding is used to fabricate CMs with varying BFS contents (0–30 wt.%) in a PP resin, both with and without 20 wt.% SGF. When it comes to its variables—materials or operational conditions—erosion is a non-linear process. The process's influencing factors must be adjusted by the desired erosion rate values to obtain them consistently and accurately. Since the number of such parameters is too large and the parameter–property correlations are not always known, statistical techniques can be utilized to precisely identify key control parameters for optimization. The erosion wear response of these BFS-filled composites is examined through a series of carefully monitored experiments that are based on the Taguchi method (L_{16} orthogonal array). The wear rate of PP resin with and without SGF was investigated. The findings showed that BFS wt.% and impact velocity are the two most significant parameters. This study attempts to generate predictive equations for the important variables that affect E_r determination (using the SYSTAT 7 software). Additionally, an ANN technique is used to estimate the composites' rate of wear. A well-trained neural network performs exceptionally well in these kinds of operations, as seen by the excellent agreement between the estimated and experimental erosion wear rate values. According to this study, adding BFS greatly increases the E_r of GF/PPCMs, making them appropriate for tribological applications.

Pradeep had developed glass fibre–polymer composites bonded with granulated BFS (GBFS) [7] using the hand lay-up process. The GBFS was ball milled to a range of 4–6 nm in four distinct wt.%: 0%, 5%, 10%, and 15%. These manufactured slabs are cut to the necessary size, and mechanical tests, including compression, tension, flexural, impact, and hardness tests, were conducted. When GBFS content went up from 0 wt.% to 15 wt.%, it was discovered that tensile, compressive, flexural strengths and hardness increased correspondingly from 242 MPa to 327 MPa, 77 MPa to 112 MPa, 45 MPa to 65 MPa, and 33.62 BHN to 139.07 BHN. In addition, impact tests (Charpy and Izod) were conducted, showing increases of roughly 23% and 100%, respectively, with 15% weight percentage filler content. Based on experimental results, it is observed that the addition of GF and GBFS to resin improves its mechanical properties. Additionally, the specimen with the highest percentage of GBFS (15%) has better tensile, impact, flexural, compression, and hardness than the other specimens.

Mostafa et al. looked at how introducing BFS as a filler affected the mechanical, thermal, and rheological characteristics of polypropylene (PP) and polystyrene (PS) [8]. Melt kneading was used to introduce 10%, 20%, and 30% weight fractions of both types, which were processed into batches of 71 μm, 40 μm, and 20 μm. Thereafter, the mixtures were crushed into plates that were 2-mm thick for thermal and rheological characteristics and 4-mm thick for tensile characteristics. When BFS filler sizes up to 71 μm and weight fractions up to 30% were used, shear viscosity increased noticeably. As filler loading increased, the crystallization enthalpy, temperature, and thermal behaviour of BFS/PP formulations decreased. The size, type, or amount of the slag micro-filler had no noticeable impact on the glass transition temperature for the PS formulations, and only very minor variations in values as compared to unfilled polymer were observed. Reports were made about the tensile characteristics of PS and PP formulations in comparison to unfilled polymers.

The ultimate strengths of the PP formulations dropped by 16% and 21% (for 40 μm ACBS and GBS), 14% and 18% (for 20 μm ACBS and GBS), and 14% and 20% (for 71 μm ACBS and GBS). The ultimate strengths of PS formulations fell by 18% and 14% (for 71 μm ACBS and GBS), 14% and 17% (for 40 μm ACBS and GBS), and 14% and 17% (for 20 μm ACBS and GBS).

The objective of Erdoğan et al. [9] was to develop an alternative use for the BFS that harms factories and the atmosphere. This led to an investigation into the suitability of slags from BFs, converters, and ferrochromium as reinforcing materials for epoxy composites used in tribology. The resin was mixed with 61-mm-sized Al_2O_3, ferrochromium slag (FeCrS), converter slag (COS), and BFS. In ball-on-disk testing, the epoxy composites were evaluated at 1,500 rpm rotating speed and 300 m sliding distance against a stationary Al_2O_3 ball with loads of 10 N, 15 N, and 20 N. Al_2O_3, BFS, COS, and FeCrS-reinforced CMs are coded as S1 to S4 for easier explanations. In the S1 composite, it is evident that the coefficient of friction (CoF) rises with the increase in load (Figure 5.1a). The S3 specimen's friction coefficient graph (Figure 5.1c) differed from the S1 at 15 N and 20 N loads, but it was identical to the S1 under light load. The friction coefficient of the S2 composite in the graph (Figure 5.1b) was similar to that of the S3, with higher values obtained under

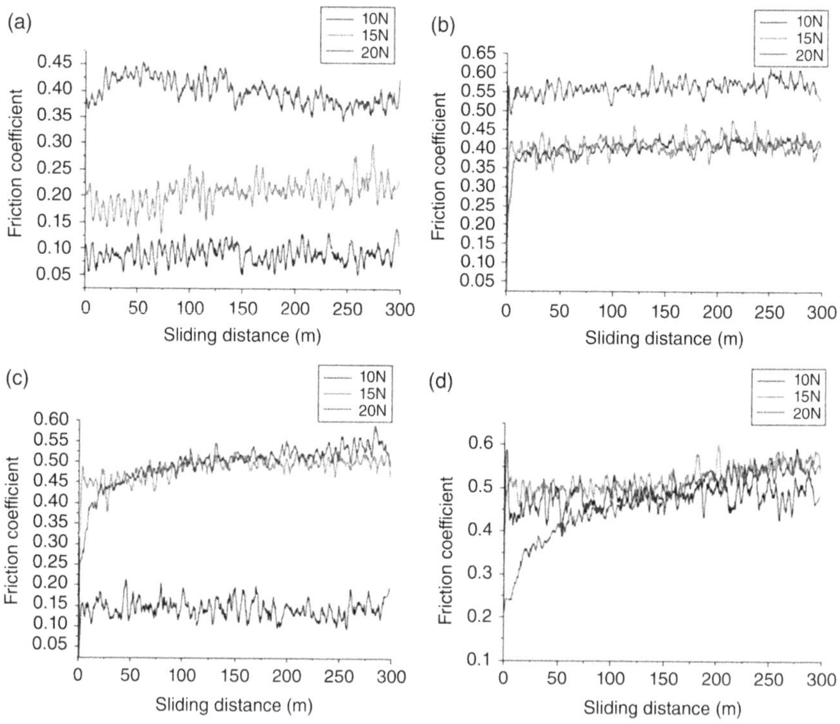

Figure 5.1 Relationship between friction coefficient and sliding distance in composite material specimen (a) S1 (b) S2 (c) S3 (d) S4 [9].

the load of 20 N and close values under the load of 10 N and 15 N. The S4 composite's friction coefficient values (Figure 5.1d) intersected at a distance of 150 m. The relative motion of the ball and the opposing part influences forces on the surface of the material, which occurs in addition to the normal load. In this study, different loads were applied to EP-based CMs filled with COS, FeCrS, BFS, and Al_2O_3. The wear behaviours of these materials were studied. The best wear resistance was shown by the S2 composite under all loads. Under 10 N, 15 N, and 20 N load, the wear resistance in increasing order are S1<S4<S3, S4<S1<S3<S2, and S4<S3<S1<S2, respectively. There was no obvious proportionate relationship between volume loss and the coefficient of thermal conduction. According to this study, it is possible to establish a cleaner environment by using industrial wastes as reinforcement material in composites.

A hybrid CM has been developed by Patnaik et al. [10]; they are made of non-woven viscose fabric that has been needle-punched, reinforced with epoxy, and filled with varying wt.% of BFS (0–15%). Mechanical examination of these composites indicates that after filler loading reaches 10 wt.%, characteristics, including tensile, inter-laminar shear, and flexural strength, start to decline. Meanwhile, hardness and impact strength increase by approximately 55% and 40%, respectively, as the filler weight percentage rises. The slurry abrasion test rig, following ASTM G105 standards, is utilized to analyse the abrasive wear characteristics of different composite materials. The results of the slurry abrasion test indicated that including BFS particle fillers into viscose fabric/epoxy composites enhanced their resistance to wear. A wear rate minimum is obtained by combining the following parameters: sand particles of 100 μm, a scratch distance of 200 m, a load of 40 N, filler content of 10 wt.%, and scratch velocity of 0.93 m/s by Taguchi analysis and parametric optimization. The research indicates that the size of the sand is the important control element, followed by sliding velocity, filler content, and sliding distance. Normal load is the least significant control factor for wear rate. These composites can be utilized in the automobile industry to prepare PC boards, seat backs, headliners, and dashboards; they can also be used in sports applications like skateboarding and to make cases for laptops and mobile phones. These composites made of fabrics have a lot of promise for the building sector and can serve as plywood substitutes. Composite materials are recommended for use in applications such as elevator buckets, chute liners, floor liners for augers in grain tanks, and offshore components.

A more efficient alternative to the environmentally harmful copper used in brake pad formulations was identified by Kanagaraj et al. [11]. Three new friction composite materials were produced by incrementally increasing the weight percentage of ground granulated BFS (GGBFS), a viable alternative to copper, from 5% to 15% in 5% increments. The other two elements, namely the combinations of inert filler and GGBFS, were changed to 5 wt.%, 10 wt.%, and 15 wt.% and 25 wt.%, 20 wt.%, and 15 wt.%, respectively, while the parent ingredients (70 wt.%) containing fibres, binders, and friction modifiers

Table 5.1 List of specimens with code

S. No.	Code	Matrix and filler(s) wt.%
1	EP	Unfilled epoxy
2	EPBFS10	Epoxy + BFS 10 wt.%
3	EPBFS10S1	Epoxy + BFS 10 wt.% + SF 1 wt.%
4	EPBFS10S3	Epoxy + BFS 10 wt.% + SF 3 wt.%
5	EPBFS10S5	Epoxy + BFS 10 wt.% + SF 5 wt.%

Source: Ref. [12].

stayed constant in composition. The created friction composites' mechanical, chemical, and physical characteristics were examined by industry requirements. Using the chase test rig, the tribological qualities were examined by the SAE J661 standard. According to preliminary research, the CMs containing 5% GGBFS had superior mechanical, chemical, and physical characteristics. It also performed exceptionally well in terms of friction, exhibiting little changes even at elevated temperatures. It is observed that the CMs with 15 wt.% GGBFS had higher resistance due to the formation of a lubricating layer at the contact, compared to the other two composites.

To manufacture an epoxy composite, Yadav et al. [12] combined sisal fibre (SF) with BFS as a hybrid filler ingredient. SFs are short, while BFS is in the shape of micro-particulates. The BFS content of the composites is maintained at 10 wt.%, while the SF content is varied between 0 wt.% and 5 wt.% (Table 5.1).

An analysis is conducted on the mechanical and physical properties of the epoxy/BFS/SF hybrid composites. Adding BFS to the epoxy matrix raises the measured density. The density of the epoxy increases from 1.13 g/cm^3 to 1.175 g/cm^3, showing a 3.98% growth. Additionally, the overall density drops with the inclusion of SF as an extra filler. Adding 5 wt.% SF reduces the density to 1.163 g/cm^3. This is a 1.02% decrease in comparison to epoxy/BFS composites. The composite absorbs water more quickly when SF or BFS are added. The pure epoxy displays a tensile strength of 26.3 MPa, whereas the cured EPBFS10 composite has a tensile strength of 33.8 MPa. There is a 28.5% rise. The tensile strength of the EPBFS10S5 composite is 40.2 MPa, showing an increase of 18.93% compared to the EPBFS10 set composite and 52.85% compared to epoxy. The epoxy's flexural strength is 36.4 MPa, and for the set EPBFS10 composite, it increases to 42.1 MPa. It is an increase of 15.66%. Flexural strength for the set EPBFS10S5 composite rises to 48.9 MPa, increasing by 34.34% compared to epoxy and 16.15% compared to EPBFS10 composite. With a 16.67% increase against epoxy, the set EPBFS10S5 has a compressive strength of 94.5 MPa. In the same way, set EPBFS10's shore-D hardness is 82 as opposed to epoxy's 80. For set EPBFS10S5 composites, the 83 Shore-D number for hardness is tested. For the highest filler content, a maximum improvement of 3.75% is observed. The currently manufactured composites have better qualities and can be used for light-duty structural applications such as floor tiles, door panels, and partition boards.

5.2.2 On utilization of red mud as filler in polymeric matrix

The caustic leaching of bauxites to make alumina results in red mud (RM) as a by-product. Roughly, 30–45% of the bauxite that is processed is wasted as red mud. This waste provides major challenges of storing and environmental degradation. Therefore, the utilization of such industrial waste is the need of the hour. Polymethylmethacrylate/red mud (PMMA/RM) and polyvinylchloride/red mud (PVC/RM) nanocomposites' thermal stabilities were examined concerning the impacts of chemical modifications on RM [13]. Thermal gravimetric analysis (TGA) and a thermal mechanical analyser (TMA) were used to examine the thermal permanence of the composites. Treatments led to the formation of basic or acidic functional groups on RM surfaces, which changed the acid–base and pH values of the red mud under study. Figure 5.2 displays the TGA analysis for the two nanocomposites: (a) RM/PMMA and (b) RM/PVC. By TGA thermograms, RM/PMMA and RM/PVC nanocomposites start to break down at 340°C and 239°C, respectively. In the figure, VRM means untreated RM, ARM means academically treated RM, and BRM means treated RM.

A novel conducting composite was developed in 2007 by employing polyaniline (PANI) and RM as the inorganic substrate [14]. The conductivities of RM/PANI composites vary from 0.39 S/cm to 4.9 S/cm depending on the weight percentage of PANI. The mechanical characteristics of the composites are also impacted by the addition of RM in a polymeric phase. When more red mud (up to 40%) was added to polypropylene, an increase in hardness values was observed. However, the tensile and flexural strength values decreased with increasing red mud addition to polypropylene [15]. On the other hand, the addition of short banana fibres (10–15 mm) to the polyester/RM hybrid composites improved the flexural, tensile, and impact strength [16, 17].

Reduced concentrations of the inorganic elements in RM exhibit dielectric characteristics, with a variety of nanoparticle sizes. Red mud has the unique ability to significantly reduce the complicated permittivity. When 0.5% and 2% are added, the polymer's real part of permittivity decreases from 5.75 to

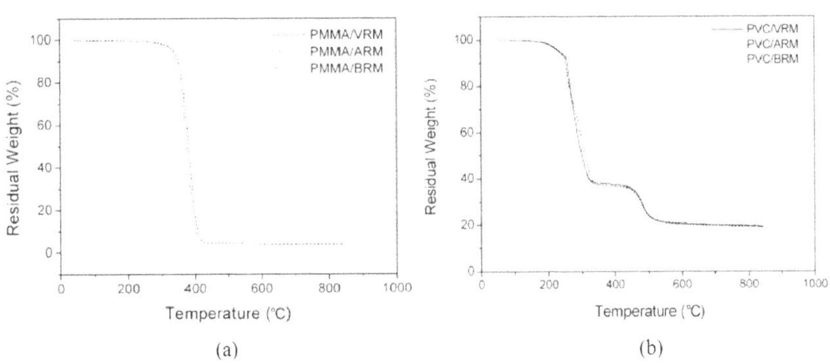

Figure 5.2 TGA curves of (a) RM/PMMA and (b) RM/PVC nanocomposites [13].

Table 5.2 Theoretical and actual densities of epoxy/RM composites

RM content (vol.%)	Measured density (g/cm³)	Theoretical density (g/cm³)	Void content (%)
5	1.16	1.2	3.33
10	1.23	1.3	5.38
15	1.32	1.4	5.71
20	1.39	1.5	7.33
25	1.48	1.6	7.50

Source: Ref. [19].

3.75 [18]. With reduced permittivity, the polymer composites can be used in areas such as antennas for the engineering of radomes, radioelectronic devices to fill the gaps between electrical circuits, etc.

RM can also be added to the thermosetting polymer like epoxy to develop composites. Table 5.2 displays the RM/EP composites' theoretical and real densities. The existence of voids causes the actual densities of the composites to be lower than their theoretical densities. Moreover, it has been shown that when RM content rises, so does the void percentage of the composites. Additionally, the addition of 25 vol.% of RM resulted in a 135% increase in the effective heat conductivity (k_{eff}) of RM/EP composites [19].

The findings of an analytical and experimental research study into polyester/coir/RM CMs resulted in enhanced tensile and compressive strength, buckling characteristics, and hardness of the composites. Hence, the polyester/coir/RM hybrid CMs are fit for application as inexpensive construction materials [20]. Adding RM to polypropylene enhances the flexural, tensile strengths, and thermal characteristics of the CMs [21]. Kucukdogan et al. [22] examined the predictive accuracy of theoretical and empirical techniques to establish the thermal conductivity coefficient (TCC) of RM/PP CMs. Furthermore, research demonstrates that using RM with hybrid structures can efficiently enhance TCC. Polyester/RM nano-CMs show significant enhancements in hardness, impact energy, and tensile strength at weight percentages of 1.0, 2.0, 3.0, and 4.0. The inclusion of RM nanoparticles enhanced the wear resistance of RM/polyester composites [23].

Figure 5.3 displays the findings from the XRD and EDAX analyses of the red mud. There are traces of several elements, including sodium, iron, carbon,

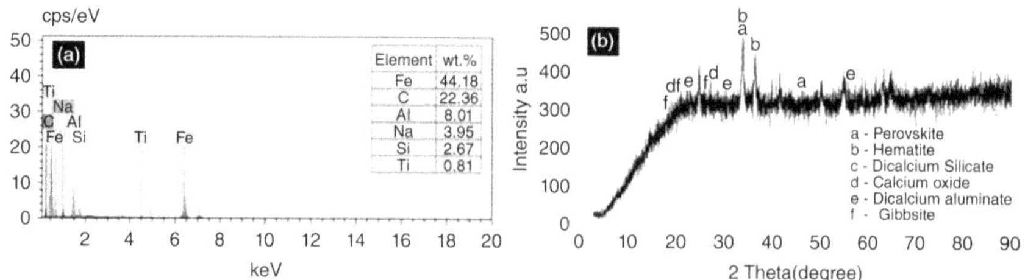

Figure 5.3 RM characterization. (a) EDAX. (b) XRD [24].

aluminium, silica, and titanium. The primary mineral constituents found in red mud are dicalcium aluminate, gibbsite, hematite, perovskite, dicalcium silicate, and calcium oxide [24].

It has been discovered that appropriately treated RM can enhance composites' electrical conductivity, damping and radiation protection qualities, mechanical, thermal, and wear properties, as well as their chemical and water resistance [25]. A recent study discovered that RM-filled polymer composites can be gainfully used as radiation-shielding components [26]. Moreover, RM can be utilized to significantly lower production costs in resin composites as a traditional filler. More significantly, the use of RM in plastic goods can lower carbon dioxide emissions and waste residue from the metallurgical sector, protecting the environment and fostering the long-term growth of the plastics industries, metallurgical, and building.

5.2.3 On utilization of LD slag and sludge as filler in polymeric matrix

Linz-Donawitz slag (LD slag) and Linz-Donawitz sludge (LDS) are among the solid wastes generated in significant amounts by steel plants. They typically include substantial amounts of valuable metals and minerals. The most important problem in the current era of material production is the generation of waste in addition to the desired material. In this context, recycling and recovering are the best ways to protect the environment from pollution. Composites can be manufactured using LD slag. Figure 5.4 displays an SEM image of crushed LD slag. The granulated slag is shown to have a varied size range. The Vickers hardness tester and the universal tensile testing machine (UTM) were utilized to assess the material's hardness and compression

Figure 5.4 SEM micrograph of LD slag [27].

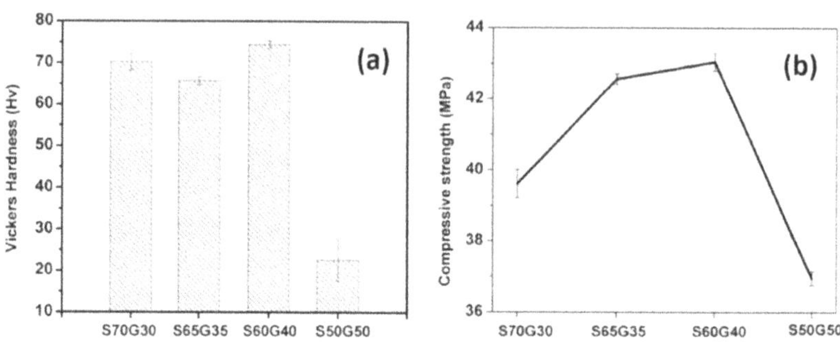

Figure 5.5 (a) Vickers hardness for composite samples. (b) Compressive strength for composite samples [27].

strength, respectively. Figure 5.5 shows the compressive strength and hardness variations of the different composite samples. The highest values of hardness and compressive strength were found to be 75 HV and 45 MPa, respectively [27]. LD slag is effectively utilized as reinforcement in polymer composites made of epoxy and polypropylene. The produced composites are analysed for their density, porosity, micro-hardness, and strength properties. Increasing the filler content improves the hardness and impact energy of the composites, but reduces the tensile and flexural properties of epoxy and polypropylene composites filled with LD slag. Adding 30 wt.% of LD slag to the epoxy matrix results in a maximum hardness of 27.94 HV and impact strength of 25.8 kJ/m². The tensile strength is 36.43 MPa and the flexural strength is 10.51 MPa [28]. The material reached a peak hardness of 79.03 HV and an impact strength of 29.7 kJ/m² by adding 30 wt.% of LD slag to a polypropylene matrix. The tensile strength was determined at 26.63 MPa and the flexural strength at 13.42 MPa [29].

LDS can serve as filler in polymer composites. The study investigates the mechanical characteristics of LDS, including industrial by-products such as BF slag and LD slag. Figures 5.6–5.8 display the changes in micro-hardness, tensile strength, and flexural strength of epoxy composites, including LDS, LD slag,

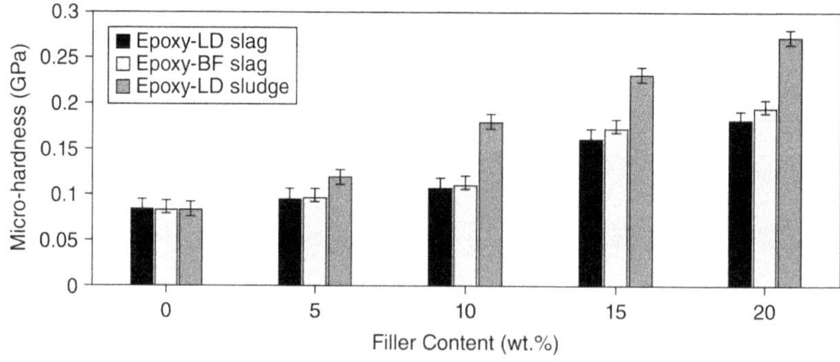

Figure 5.6 Comparison of micro-hardness of LD sludge-, BF slag-, and LD slag-filled composites [30].

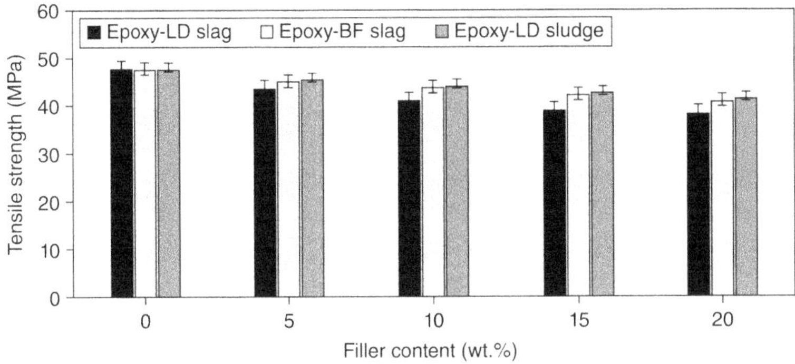

Figure 5.7 Comparison of tensile strength of LD sludge-, BF slag-, and LD slag-filled composites [30].

and BF slag, respectively. The inclusion of filler improved the micro-hardness of the composite. By incorporating 5 wt.% of LDS into the pure epoxy, the micro-hardness increased from 0.085 GPa to 0.12 GPa. Micro-hardness values increased to 0.181 GPa, 0.232 GPa, and 0.273 GPa with the addition of 10%, 15%, and 20% LDS to pure epoxy, respectively. With an increase in filler content, the tensile and flexural strength of the composite decreased gradually. Comparatively, epoxy filled with LD sludge has superior mechanical properties among the three fillers [30]. The physical and mechanical characteristics of PP composites reinforced with LDS were analysed. As the LDS filler loading rises, density also increases. The neat PP has a density of 0.8998 g/cm³.

The density rises to 0.9172 g/cm³ with the addition of 5 wt.% filler and further increases to 0.9744 g/cm³ with 20 wt.% filler. Adding LDS significantly affects compressive strength and hardness. Both compressive strength and hardness increase when LDS filler is added to the PP matrix. The compressive strength of unfilled PP is 82 MPa and its hardness is 0.058 GPa. When the LDS

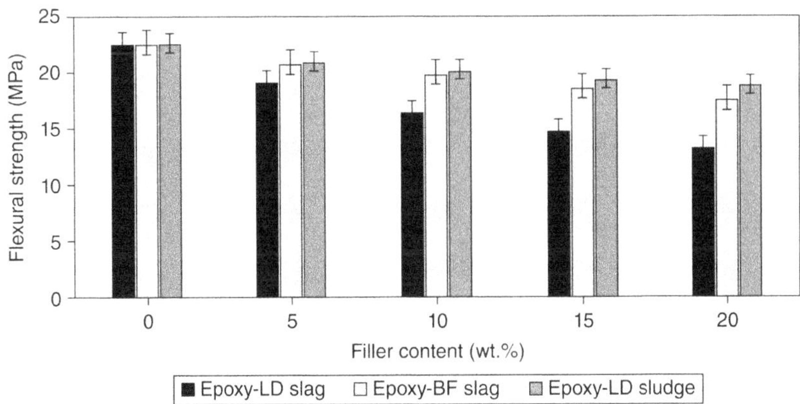

Figure 5.8 Comparison of flexural strength of LD sludge-, BF slag-, and LD slag-filled composites [30].

content reaches 20 wt.%, the compressive strength and hardness peak at 85.71 MPa and 0.643 GPa, respectively. Tensile and flexural strength decrease as filler loading increases. When 5 wt.% of filler is added, the tensile strength reduces slightly to 38.28 MPa. When the filler concentration reaches 20 wt.%, there is a significant decrease in the tensile strength. The tensile strength decreases to 33.94 MPa. The unfilled PP has a flexural strength of 29.23 MPa. Flexural strength diminishes with higher filler content in the PP. The flexural strength of PP filled with 5% weight of LDS is 28.15 MPa but decreases to 22.98 MPa when the filler percentage is increased to 20% weight [31]. An increase in LDS content leads to higher density and hardness in epoxy-based composites but results in a decrease in tensile and flexural strength. The theoretical density of the composites increases from 1.1 g/cm^3 for pure epoxy to 1.162 g/cm^3 when 20 wt.% LDS is added. The epoxy composite's hardness is 0.085 GPa, and it rises to 0.273 GPa when 20 wt.% LDS material is added. Pure epoxy has a tensile strength of 48 MPa, which reduces to 41.6 MPa when 20 wt.% LDS content is added. The epoxy's bending strength is 22.5 MPa; however, it reduces to 20.9 MPa when 5 wt.% of LDS is added. The flexural strength of the composites reduces to 20.1 MPa, 19.3 MPa, and 18.8 MPa when the compositions contain 10 wt.%, 15 wt.%, and 20 wt.% of LDS, respectively [32].

5.2.4 On utilization of copper slag as filler in polymeric matrix

Copper slag is an oxide-based industrial waste generated during the production of copper from ore by pyrometallurgical process [33]. The raw copper slag typically contains the compounds of aluminium, silicon, magnesium, calcium, and majorly iron [34]. The chemical composition of copper slag ensures the presence of about 55% Fe_2O_3, 31% SiO_2, 2.5% Al_2O_3, 1.5% MgO, 5% CaO, 0.25% Na_2O, 2% SO_3, 0.6% K_2O, and 1.7% of LOI in it [35]. The photographic view and the scanning electron micrograph of ball-milled copper slag are displayed in Figures 5.9a and 5.9b, respectively. The micrograph displays that the crushed slag particles are irregularly shaped and consist of sharp edges.

(a)

(b)

Figure 5.9 (a) Photographic image of copper slag [36]. (b) SEM of copper slag particles [37].

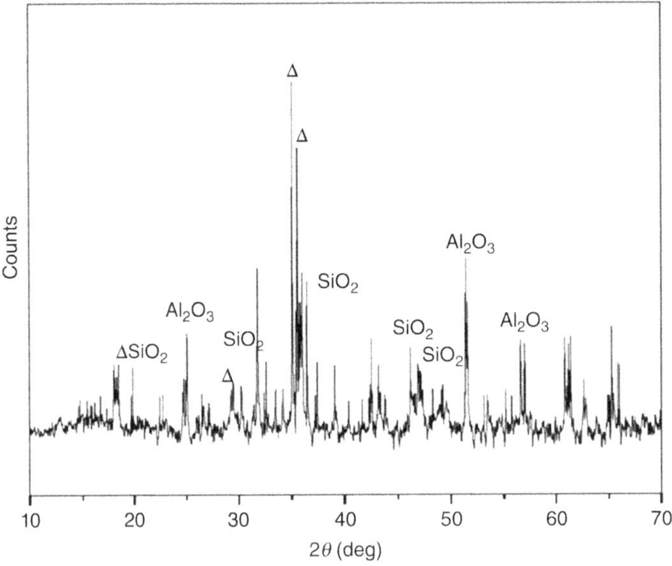

Figure 5.10 XRD of copper slag particles [39].

To ascertain the compounds present in copper slag and the organic phases that exist within the elements present in it, typical X-ray diffractograms and Fourier transform infrared spectroscopy (FTIR) of raw copper slag are presented in Figures 5.10 and 5.11, respectively. The X-ray diffractograms ensure that iron oxide (Fe_2O_3), silicon oxide (SiO_2), aluminium oxide (Al_2O_3), and magnesium oxide (MgO) are the major constituents of copper slag as the

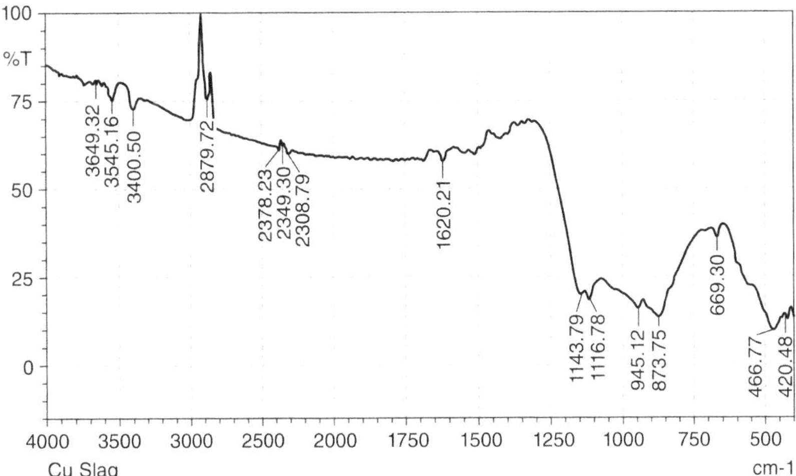

Figure 5.11 FTIR spectroscopy of copper slag [40].

larger diffraction peaks represent these compounds. The other compounds like calcium oxide (CaO), sodium oxide (Na_2O), potassium oxide (K_2O), sulphur trioxide (SO_3), and so on are the minor constituents of raw copper slag as very few diffraction peaks having smaller intensity represent these compounds. The FTIR spectroscopy of raw copper slag taken in transmission mode ensures the presence of the OH group in bound water, the asymmetric stretching of the C–O bond of the $CO_3{}^{2-}$ group in calcite ($CaCO_3$), and the asymmetric stretching vibration of S–O bonds of $SO_4{}^{2-}$ groups in it [38].

Erdenebold et al. [34] in their research reported that about 2.2–3 tons of raw copper slag is generated per ton of pure copper. Thus, the increasing production of copper multiplies the slag production day by day. The statistics prepared by the Ministry of Mines (India) reported that the average copper production per year (accounting last 5 years) is about 93,000 metric tons [41]. Thus, the amount of slag produced should be about 230,000 metric tons per annum only in India. Being rich in so many metal oxides, copper slag has the potential to be used in numerous applications but still its reusable potential has not been fully exploited yet. However, in previous few decades, it has attracted the attention of potential researchers to make its utilization as a potential filler in the preparation of concrete [35], as the feedstock in the deposition of coatings [42], and also as a filler in the fabrication of wear-resistant polymer composites [40, 43].

The past studies hence showed that copper slag, instead of being an industrial waste, holds the reinforcing potential to be used for the fabrication of polymer-based composites. In a recent research, Barczewski et al. [44] used copper slag as a particulate filler for the preparation of polylactide (PLA)-based composite and studied the mechanical and thermomechanical properties-based changes of the composites corresponding to the filler concentration. They concluded that the mechanical strength properties of the composites were positively affected by the increase in filler amount by up to 10% by weight. Up to this amount, the filler particles were properly distributed within the matrix body and no agglomeration was identified. An SEM image of the composite explaining the proper bonding between the filler and matrix is displayed in Figure 5.12. The figure shows proper wetting of the filler particles within the matrix and almost negligible voids are present within it.

Kalusuraman et al. [40] elucidated the effect of copper slag on the erosion response of jute–polyester composites. The study found that adding copper slag as a secondary filler up to 10 wt.% in the jute–polyester composite improved the bonding between the fibres and matrix, resulting in increased resistance to erosion wear. The erosion rate was shown to rise with jet velocity but decrease with impingement angle. Using copper slag as a filler in polyethylene-based composites decreased the composite's tensile strength by approximately 5%. A 16% increase in the tensile modulus was observed with a 5 wt.% filler addition [45]. Biswas et al. [46] investigated the impact of copper slag on the physical and mechanical characteristics of bamboo-fibre-reinforced epoxy composites. The density of the fibre–polymer composites increased from 1.15 g/cm^3 to 1.38 g/cm^3 when the filler content increased from 0% to 15% by weight.

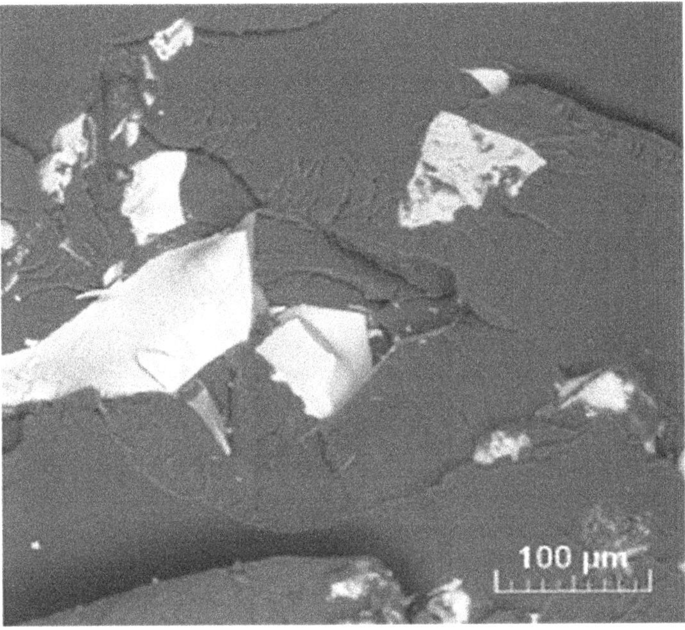

Figure 5.12 SEM image of PLA composite containing 10% copper slag [44].

Nevertheless, the void fractions were also raised when the filler was added. Significant increases in the hardness and tensile strength values of the composites were noted. The flexural and impact strength of the composites exhibited a non-linear trend, decreasing with filler addition up to 5 wt.% and then increasing thereafter. Adding micro-sized copper slag as a supplementary filler in the fibre–polymer composite system enhanced the physical and mechanical strength qualities of the composites. Biswas and Satapathy [43] studied the impact of copper slag on the mechanical and erosion wear parameters of glass fibre-reinforced epoxy composites. Copper slag can be used as an auxiliary filler in the glass–epoxy composite system up to 20% by weight, according to their findings. By including this waste as a filler, a little decrease in the tensile, flexural, and inter-laminar shear strengths has been noted. The composites' impact strength and micro-hardness improved as the filler quantity increased. The glass–epoxy composite showed a notable increase in resistance to erosion wear, regardless of the other control parameters.

Considering the outcomes reported in the previous research using copper slag as a potential filler in the fabrication of polymer composites, including thermosetting and thermoplastic ones, copper slag is well-justified to be used as a suitable reinforcing agent. The research on copper slag is limited compared to other industrial wastes like red mud and BF slag, despite its ability to enhance the physical, mechanical, and tribological properties of polymers. There is significant potential for further research utilizing copper slag.

5.3 CONCLUSIONS

It has been observed from past research that industrial waste in the form of filler material has emerged as a potential filler material for the development of polymer matrix composite material. Keeping in mind, the current discussion has been intended to deliver a review with a main focus on the composites prepared with polymer as a base matrix and industrial waste in the form of microparticulates as a filler material. The industrial waste discussed in this chapter is the waste generated from iron and steel industries, aluminium industries, and copper industries. It has been discussed, on the basis of findings available, how the inclusion of industrial waste influences the various properties of polymers of different kinds. The properties of the composites discussed in this chapter are physical, mechanical, thermal, and tribological properties. These composites are suitable for many applications such as partition boards, false ceilings, pipelines for coal dust, chute liners, pulley laggings, exhaust fan blades, nozzles, diffusers, and lightweight vehicles. These materials can also be recommended for use in engineering buildings exposed to dusty environments and as affordable building materials in desert areas. Therefore, it can be inferred that these composites demonstrate encouraging outcomes, endorsing their suitability for diverse light-duty applications on a larger scale in the foreseeable future.

5.4 LIMITATIONS AND FUTURE RESEARCH DIRECTIONS

The compatibility of industrial waste fillers with different polymer matrices may vary, leading to challenges in achieving desired physical, mechanical, tribological, and thermal properties. The chapter may discuss specific processing techniques for incorporating industrial waste fillers into polymer composites, but limitations related to scalability, cost-effectiveness, and energy consumption should be acknowledged. Long-term durability and environmental stability of polymer composites containing industrial waste fillers need further investigation to assess their performance under various conditions. The potential leaching of contaminants from industrial waste fillers into the environment and their impact on human health and ecosystems should be studied comprehensively. Further, the specific applications such as automotive components, building materials, or medical devices to understand the performance of polymer composites with industrial waste fillers in different contexts can be explored.

REFERENCES

1. Murari K, Siddique R, Jain KK. Use of waste copper slag, a sustainable material. J Mater Cycles Waste Manag 2015;17(1):13–26.
2. Shi C, Qian J. High-performance cementing materials from industrial slags: A review. Resour Conserv Recycl 2000;29:195–207.

3. Escalante-Garcia JI, Espinoza-Perez LJ, Gorokhovsky A, et al. Coarse blast furnace slag as a cementitious material, comparative study as a partial replacement of Portland cement and as an alkali activated cement. Constr Build Mater 2009;23:2511–2517.

4. Padhi PK, Satapathy A. Prediction and simulation of erosion wear behavior of glass- epoxy composites filled with blast furnace slag. Adv Mater Res 2012;585:549–553.

5. Padhi PK, Satapathy A. Analysis of sliding wear characteristics of BFS filled composites using an experimental design approach integrated with ANN. Tribol Trans 2013;56:789–796.

6. Padhi PK, Satapathy A. Processing and wear analysis of blast furnace slag filled polypropylene composites using Taguchi model and ANN. Int Polym Process 2014;29:233–244.

7. Pradeep AV. Effect of blast furnace slag on mechanical properties of glass fiber polymer composites. Procedia Mater Sci 2015;10:230–237.

8. Mostafa A, Laske S, Pacher G, et al. Blast furnace slags as functional fillers on rheological, thermal, and mechanical behavior of thermoplastics. J Appl Polym Sci 2016;133:1–8.

9. Erdoğan A, Gök MS, Koç V, et al. Friction and wear behavior of epoxy composite filled with industrial wastes. J Clean Prod. 2019;237:117588.

10. Patnaik PK, Swain PTR, Biswas S. Investigation of mechanical and abrasive wear behavior of blast furnace slag-filled needle-punched nonwoven viscose fabric epoxy hybrid composites. Polym Compos 2019;40:2335–2345.

11. Kanagaraj M, Babu S, Raj S, et al. Influence of ground granulated blast furnace slag on the tribological characteristics of automotive brake friction materials. Ind Lubr Tribol 2022;74:837–843.

12. Yadav P, Agrawal A, Ayachit B, et al. Physical and mechanical behaviour of epoxy based hybrid composites with blast furnace slag and sisal fiber as reinforcement. Mater Today Proc 2022;63:268–271.

13. Park SJ, Jun BR. Improvement of red mud polymer-matrix nanocomposites by red mud surface treatment. J Colloid Interface Sci 2005;284:204–209.

14. Gök A, Omastová M, Prokeš J Synthesis and characterization of red mud/polyaniline composites: Electrical properties and thermal stability. Eur Polym J 2007;43:2471–2480.

15. Akinci A, Akbulut H, Yimaz E (2008) Mechanical properties of cost-effective polypropylene composites filled with red-mud particles. Polym Compos 2008;16:439–446.

16. Prabu VA, Manikandan V, Uthayakumar M. Effect of red mud on the mechanical properties of banana/polyester composites using design of experiments. Proc Inst Mech Eng Part L: J Mater Des Appl 2013;227:143–155.

17. Arumuga Prabu V, Uthayakumar M, Manikandan V, et al. Influence of red mud on the mechanical, damping and chemical resistance properties of banana/polyester hybrid composites. Mater Des 2014;64:270–279.

18. Bhat AH, Khalil HPSA, Haq Bhat I, Banthia AK. Dielectric and material properties of poly(vinyl alcohol)-based modified red mud polymer nanocomposites. J Polym Environ 2012;20:395–403.

19. Banjare J, Sahu YK, Agrawal A, Satapathy A. Physical and thermal characterization of red mud reinforced epoxy composites: An experimental investigation. Procedia Mater Sci 2014;5:755–763.

20. Rachchh NV, Misra RK, Roychowdhary DG Effect of red mud filler on mechanical and buckling characteristics of coir fibre-reinforced polymer composite. Iran Polym J 2015;24:253–265.

21. Zhang Y, Zhang A, Zhen Z, et al. Red mud/polypropylene composite with mechanical and thermal properties. J Compos Mater 2011;45:2811–2816.

22. Kucukdogan N, Aydin L, Sutcu M. Theoretical and empirical thermal conductivity models of red mud filled polymer composites. Thermochim Acta 2018;665:76–84.

23. Suresh S, Sudhakara D. Investigation of mechanical and tribological properties of red mud-reinforced particulate polymer composite. J Bio- Tribo-Corrosion 2019;5:1–8.

24. Vigneshwaran S, Uthayakumar M, Arumugaprabu V. Potential use of industrial waste-red mud in developing hybrid composites: A waste management approach. J Clean Prod 2020;276:124278.

25. Wu P, Liu X, Zhang Z, Wei C. Properties of red mud-filled and modified resin composites. Constr Build Mater 2023;409:133984.

26. Shivani, Vishwakarma J, Dhand C, et al. "Red-mud, a golden waste for radiation shielding": Red-mud polymer composites for high-performance radiation-shielding components. J Hazard Mater Adv 2024;13:100394.

27. Mallik M, Hembram S, Swain D, Behera G, Potential utilization of LD slag and waste glass in composite production. Mater Today Proc 2020;33:5196–5199.

28. Pati PR, Satapathy A. A study on processing, characterization and erosion wear response of Linz-Donawitz slag filled epoxy composites. Adv Polym Technol 2015;34(4):21509, 2015.

29. Pati PR, Satapathy A. Processing, characterization and erosion wear response of Linz-Donawitz (LD) slag filled polypropylene composites. J Thermoplast Compos Mater 2016;29(9):1282–1296.

30. Purohit A, Satapathy A. Mechanical and wear characteristics of epoxy composites filled with industrial wastes: A comparative study. IOP Conf Ser: Mater Sci Eng 2017;178:012019.

31. Purohit A, Gupta G, Pradhan P, Agrawal A. Development and erosion wear analysis of polypropylene/Linz-Donawitz sludge composites. Polym Compos 2023;44:6556–6565.

32. Purohit A, Satapathy A. Processing, characterization, and parametric analysis of erosion behavior of epoxy-LD sludge composites using Taguchi technique and response surface method. Polym Compos 2018;39:E2283–E2297.

33. Gorai B, Jana RK, Premchand. Characteristics and utilisation of copper slag: A review. Resour Conserv Recycl 2003;39:299–313. https://doi.org/10.1016/S0921-3449(02)00171-4.

34. Erdenebold U, Choi HM, Wang JP. Recovery of pig iron from copper smelting slag by reduction smelting. Arch Metall Mater 2018;63:1793–1798.

35. Rathanasalam V, Perumalsami J, Jayakumar K. Mechanical and microstructural properties of copper slag based blended geopolymer concrete. Medziagotyra 2021;27:302–307.

36. Abdalla TA, Alahmari TS. Mechanical strength and microstructure properties of concrete incorporating copper slag as fine aggregate: A state-of-the-art review. Adv Civil Eng 2024;4389616.

37. Sharma R, Khan RA. Influence of copper slag and metakaolin on the durability of self-compacting concrete. J Clean Prod 2018;171:1171–86.

38. Feng Y, Kero J, Yang Q, Chen Q, Engström F, Samuelsson C, et al. Mechanical activation of granulated copper slag and its influence on hydration heat and compressive strength of blended cement. Materials (Basel) 2019;12:772.

39. Mantry S, Jha BB, Satapathy A. Evaluation and characterization of plasma sprayed Cu slag-Al composite coatings on metal substrates. J Coatings 2013;2013:1–7.

40. Kalusuraman G, Thirumalai Kumaran S, Aslan M, Küçükömeroğluc T, Siva I. Use of waste copper slag filled jute fiber reinforced composites for effective erosion prevention. Measurement 2019;148:106950.

41. Jaganmohan M. Production volume of copper concentrate in India from financial year 2012 to 2021, with estimates until 2023 (in 1,000 metric tons). 2024:1.

42. Mantry S, Behera D, Satapathy A, Jha BB, Mishra BK, Mantry S, et al. Deposition of plasma sprayed copper slag coatings on metal substrates. Surf Eng 2013;29:222–228.
43. Biswas S, Satapathy A. Use of copper slag in glass-epoxy composites for improved wear resistance. Waste Manag Res 2010;28:615–625.
44. Barczewski M, Hejna A, Aniśko J, Andrzejewski J, Piasecki A, Mysiukiewicz O, et al. Rotational molding of polylactide (PLA) composites filled with copper slag as a waste filler from metallurgical industry. Polym Test 2022;106:107449.
45. Hejna A, Kosmela P, Barczewski M, Mysiukiewicz O, Piascki A. Copper slag as a potential waste filler for polyethylene-based composites manufacturing. Tanzania J Sci 2021;47:405–420.
46. Biswas S, Patnaik A, Kaundal R. Effect of red mud and copper slag particles on physical and mechanical properties of bamboo-fiber-reinforced epoxy composites. Adv Mech Eng 2012;2012:1–6.

Chapter 6

Finite element modeling and buckling behaviour analysis of sandwich composite panel

Ravi Kumar, Sandeep Tiwari, Rajesh Kumar, and Chetan Kumar Hirwani

6.1 INTRODUCTION

In today's advanced engineering fields like aeronautics, automotive, and marine industries, minimizing total weight is a significant focus alongside ensuring structural strength. This factor directly impacts the operational cost and efficiency of both the structure and its associated systems. Furthermore, advanced engineering materials fulfil their primary criteria, namely, high strength, stiffness, and reduced weight. As a result, structural components made from fibre-reinforced composites and sandwich materials are experiencing widespread acceptance due to their advantages over traditionally used metallic materials [1]. Typically, a sandwich structure is created by sandwiching a thick core layer of high density between two robust thin face skins [2]. This construction results in a final structure with minimal total density (on average), characterized by a high stiffness-to-weight ratio and bending strength-to-weight ratio. These structural components are continuously utilized in specific work environments, subjected to various loads and experiencing variable degrees of deformation due to dynamic loading. Buckling stands out as a crucial failure mechanism in these conditions. The buckling phenomenon involves an abrupt shift in the equilibrium configuration at a specific critical load. Hence, it becomes imperative to analyse sandwich structures under diverse operating environments and loading conditions, demanding additional attention to ensure their performance and durability.

The buckling of composite plates is a very complicated subject about which several studies are available in open literature. Nevertheless, the amount of information obtained from available literature on composite plates is not enough and considerably more is needed. Di Sciuva [3] proposed a piece-wise displacement model to investigate the buckling, fundamental frequency and bending of symmetrical laminated simply supported orthotropic flat panels. The proposed model is also compared with the results of the exact elasticity, Kirchhoff, and shear deformation model to establish the efficacy and accuracy of the model. Semi-numerical finite strip approach was presented to analyse the buckling responses of laminated composite flat panel subjected to in-plane compression by Chai and Khong [4] and conferred the insight of

DOI: 10.1201/9781003564355-6

varying the end condition on the buckling responses and finally discussed the limitation of the proposed approach. Further, Khdeir and Reddy [5] developed an analytical solution using the state space approach in association with Jordan canonical form to evaluate the buckling behaviour of a cross-ply orthotropic rectangular beam. The results were evaluated for different geometrical parameters and end conditions. A combined single-layer and layerwise formulation was developed by Gilat et al. [6] to predict the buckling loads of homogeneous and orthotropic layered composite plates. Comparison of results with the exact solution and discrete layer theory established the accuracy of the formulated model with reduced computational cost. The effectiveness of three different theories (classical plate theory [CLT], first-order and second-order shear deformation (FOSD) plate theory) was examined by Singh et al. [7] to predict the buckling of layered composite flat panels. To better forecast the composite behaviour, they modeled the lamina properties as a random variable using a first-order perturbation technique. Furthermore, the critical buckling of layered composite flat panels subjected to uniaxial and biaxial loading was estimated by Shukla et al. [8] using FOSD theory. They found that the material properties, lay-up arrangement, and aspect ratio have significant effect on the buckling characteristics of plate. Ungbhakorn and Singhatanadgid [9]extended the Kantorovich method to estimate the buckling responses of layered composite rectangular plates under varied end conditions. The potential energy method accompanied by a separable displacement function was used to derive the governing equation for buckling. Gaussian and Multiquadric function meshless methods were adopted in the work of Singh et al. [10] for investigating the buckling behaviour of composite laminate under thermomechanical loads. The analytical meshless method was again used in a different study presented by Neves and Ferreira [11] for natural vibration and buckling analysis of layered composite plates. The novelty of this method was the practice of oscillatory radial basis function. Another analytical method based on FOSD theory in conjunction with the state space approach was presented in the work of Xing and Xiang [12] for the buckling analysis of rectangular cross-ply layered composite flat panels. Singh and Singh [13] developed new shear deformation theories based on the shear–strain interpolation function implemented to investigate the modal and buckling responses of layered and three-dimensional interweaved composite flat panels. The accuracy of the novel theories (trigonometric deformation theory and trigonometric hyperbolic deformation theory) was established by comparing results with 3D elasticity and other available numerical methods. Pre-buckling stress field effect was introduced in the analytical buckling solution of layered composite flat panel presented by Nima and Ganesan [14]. They used the Ritz approximate solution method to obtain the eigen buckling of the plate. Another work implementing a semi-analytical approach for local buckling analysis of composite thin-walled flat and curved sections was presented by Higginson et al. [15]. They adopted the Rayleigh–Ritz energy method to get the analytical buckling solutions. Chen et al. [16] employed the

Galerkin method, utilizing the first-order shear deformation theory (FSDT), to examine the stability characteristics of a sandwich structure with a corrugated core. The analysis was performed considering simply supported edge conditions. Ye et al. [17] focused their investigation on the stability analysis of a layered sandwich plate incorporating functionally graded material. They utilized the layer-wise theory to examine and evaluate the plate's stability under different conditions. Sarafraz et al. [18] investigated the natural frequency and dynamic stability (uniaxial and biaxial) responses of a sandwich flat panel having a honeycomb auxetic core by utilizing sinusoidal shear deformation theory. Xinyi et al. [19] focused on examining the buckling behaviour of a sandwich structure having a star-shaped honeycomb core by using the downscaling model. Qian et al. [20] studied stability analysis of flat sandwich structures by applying higher-order theory and the general quadrature method. Bui et al. [21] proposed a lower-order theory for studying the buckling and vibration responses of a sandwich beam (I-section). Using the power law model, Ebrahimi et al. [22] examined the buckling responses of FG sandwich structures having magnetostrictive abilities.

Based on the aforementioned literature review, it is evident that the predominant research focuses on analysing the buckling responses of layered composite flat panels, while sandwich structures receive comparatively limited attention. A noticeable gap exists in the current literature concerning the buckling behaviour of sandwich composite plates, particularly when employing the FSDT under uniaxial mechanical loading conditions. To bridge this gap, a comprehensive numerical model has been developed using the ABAQUS simulation tool. This model incorporates lower-order kinematics principles and is specifically tailored for analysing sandwich composite plates. The validity of the current model has been established through a detailed validation study, comparing predicted results with published data. With successful validation, the developed numerical model is employed to conduct detailed investigations into the buckling behaviour of flat sandwich panels. These investigations take into account various geometrical and mechanical parameters, including the aspect ratio, thickness ratio of core–face, and orthotropic ratio.

6.2 GEOMETRY DESCRIPTION AND MATHEMATICAL FORMULATION

As illustrated in Figure 6.1, the sandwich composite flat panel is a form of material assembly consisting of two outer layers referred to as a face sheet, with a core 'C' material positioned between them. The dimensions of the sandwich composite panel are denoted as l, b, and h, where l represents the panel's length, b signifies the panel's width, and h denotes a uniform panel's thickness. The thickness of the face sheet and core is indicated as h_f and h_c, respectively.

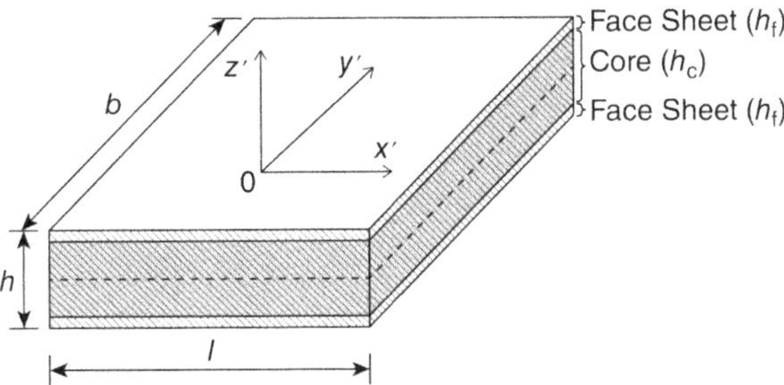

Figure 6.1 Geometry of Sandwich Composite flat panel.

6.2.1 Displacement field

Within the FSDT the assumption is made that in-plane displacements demonstrate a linear variation throughout the thickness of a composite structure. Simultaneously, the out-of-plane displacement is considered to be solely dependent on the in-plane coordinates. The mathematical expression representing the displacement field is given by

$$\left.\begin{aligned}
u(x',y',z') &= \bar{u}(x',y') + z'\varphi_{x'}(x',y') \\
v(x',y',z') &= \bar{v}(x',y') + z'\varphi_{y'}(x',y') \\
w(x',y',z') &= \bar{w}(x',y') + z'\varphi_{z'}(x',y')
\end{aligned}\right\} \tag{6.1}$$

Here, $\bar{u}$ and $\bar{v}$ represent in-plane displacements, typically aligned along the x' and y' axes, while $\bar{w}$ represents the out-of-plane displacement. The axes of rotation within the midplane are denoted by the symbols $\varphi_{x'}$ and $\varphi_{y'}$.

6.2.2 Constitutive relation

The constitutive relation for the k_{th} lamina, depicting the stress–strain relationship at any arbitrary angle θ about any arbitrary axis, is formulated as follows:

$$\{\sigma\} = \left[\bar{Q}\right]\{\varepsilon\} \tag{6.2}$$

Here the symbols $\{\sigma\}, \left[\bar{Q}\right],$ and $\{\varepsilon\}$ are employed to denote stress, reduced transformation stiffness, and strain, respectively.

6.2.3 Finite element formulation

The finite element method (FEM) is a powerful numerical technique employed in engineering and applied mathematics for solving complex physical problems. It involves dividing a structure or system into smaller, simpler elements, allowing for the representation of intricate geometries and material properties. In this ongoing study, discretization is achieved using an eight-noded (S8R) isoparametric element, each with six degrees of freedom (DOFs) per node. The displacement vector, denoted as δ, can be expressed at any point within the mid-plane as follows [23]:

$$\{\delta\} = \sum_{i=1}^{n} [N]\{\delta_i\} \tag{6.3}$$

Here $\{\delta_i\} = \{\bar{u}, \bar{v}, \bar{w}, \varphi_{x'}, \varphi_{y'}, \varphi_{z'}\}$ is denoted as the nodal displacement vector, and $[N]$ denotes the nodal interpolation function.

6.2.4 Governing equation

The ultimate expression of the governing equation for analysing the buckling of a sandwich flat panel is outlined below:

$$\delta\Pi = 0 \tag{6.4}$$

Here, $\Pi = U + U_{in}$

$$[K] - \lambda_{\bar{N}}[K_G] = 0 \tag{6.5}$$

Here, K represents the global stiffness matrix, K_G is the geometrical stiffness matrix, and $\lambda_{\bar{N}}$ signifies the critical buckling load or eigenvalue. Subsequently, the clamped end condition is incorporated into Equation (6.5) for resolution, leading to the determination of the eigenvalue or critical buckling load.

$$\bar{u} = \bar{v} = \bar{w} = \varphi_{x'} = \varphi_{y'} = \varphi_{z'} = 0$$

Clamped condition: at $x = 0$ and l

6.3 RESULTS AND DISCUSSION

In this research work, the mechanical buckling responses of a square sandwich structure were obtained numerically using the lower-order method and the commercial finite element package ABAQUS. A uniaxial mechanical force was exerted on one end of the sandwich structure, while the other end was constrained by clamped conditions, as depicted in Figure 6.2. A convergence test was conducted on the proposed numerical model by solving various numerical problems under several mesh refinements. Additionally, the

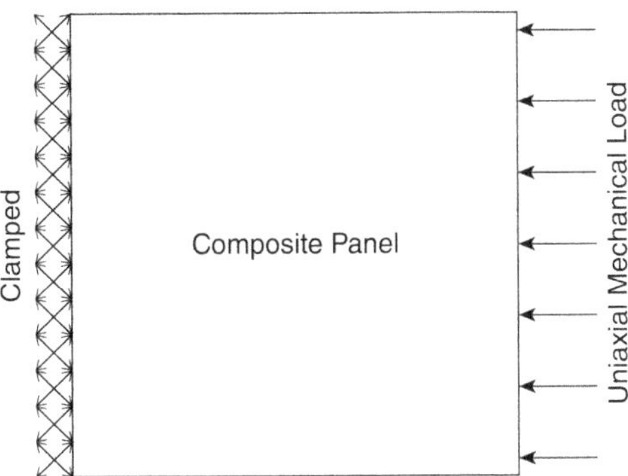

Figure 6.2 Buckling analysis for uniaxial mechanical load with one end clamped.

results obtained from the current models were compared with published data for validation purposes. Finally, various numerical examples were tackled to illustrate how the buckling behaviour of sandwich flat structures is influenced by both geometric and mechanical properties.

6.3.1 Convergence study

This section illustrates the convergence behaviour and accuracy of the current finite element (FE) model through the resolution of a numerical example. Specifically, a square (with length to breadth ratio, $l/b = 1$) sandwich plate (oriented as 0°/45°/C/45°/0°) with a thickness ratio of $(l/h) = 60$ and a core–face thickness ratio of $h_c/h_f = 12$, clamped at one end, is examined. The material properties utilized are those listed in Table 6.2. The calculated eigenvalues are transformed into non-dimensional form using Equation (6.6), and the resulting responses are depicted in Figure 6.3. It is observed that there is minimal variation in the responses after employing a (6×6) mesh refinement. Hence, this mesh resolution is adopted for subsequent analysis to determine the critical buckling load for the sandwich panel.

6.3.2 Validation study

Following the successful completion of the convergence study, the proposed simulation model is validated for buckling load parameter of symmetric layered cross-ply composite plate with thickness ratio $(l/h = 10)$ and simply supported end condition. Dimensionless buckling load parameters have been evaluated for three-layered and five-layered symmetric cross-ply

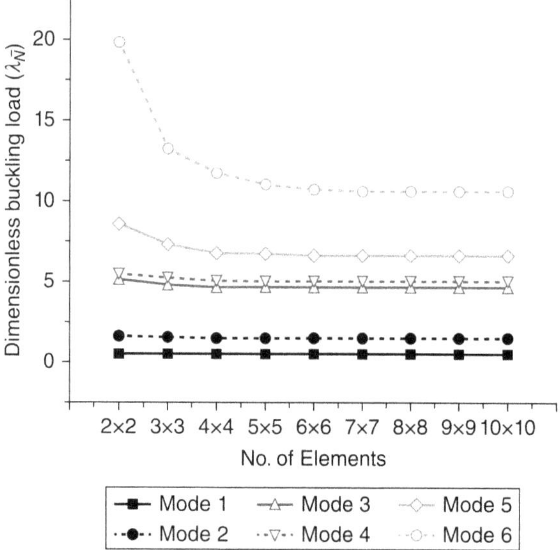

Figure 6.3 Convergence study of buckling behavior of square sandwich composite panel.

laminates and tabulated in Table 6.1. Dimensionless buckling load parameters and material properties are used the same, as given in Equation (6.6). The results are compared with the results of various theories reported in the literature by Singh and Singh [13], Putcha and Reddy [24], Noor and Burton [25], and Ferreira et al. [26]. The comparison shows that the proposed simulation model can predict the buckling response of laminated composite plates efficiently.

Table 6.1 Buckling load parameter ($\lambda_{\bar{N}}$) of simply supported symmetric ply laminates for $a/h = 10$.

		Orthotropy ratio (E_1/E_2)		
No. of layers (N)	*Source*	*3*	*10*	*20*
Three (3)	Present	5.87	9.12	12.45
	Singh and Singh [13]	5.40	9.87	14.99
	Putcha and Reddy [24]	5.39	9.94	15.29
	Noor and Burton [25]	5.30	9.76	15.01
	Ferreira et al. [26]	5.38	9.83	14.89
Five (5)	Present	6.00	9.54	13.52
	Singh and Singh [13]	5.41	10.11	15.84
	Putcha and Reddy [24]	5.40	10.15	16.00
	Noor and Burton [25]	5.32	9.96	15.65
	Ferreira et al. [26]	5.40	10.08	15.79

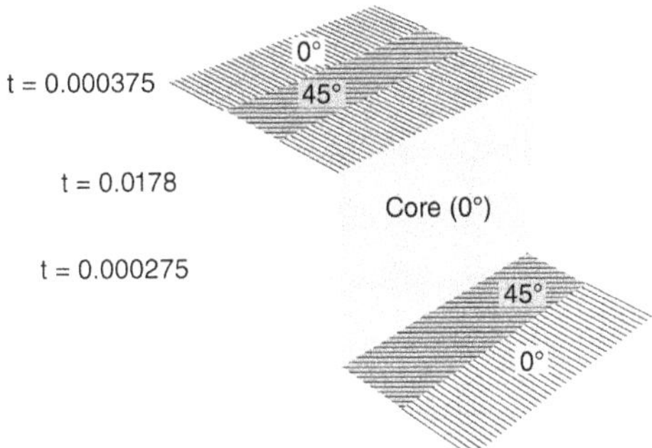

Figure 6.4 Layup scheme in sandwich composite.

Material properties:

$$\frac{E_1}{E_2} = \text{variable} \quad G_{12} = G_{13} = 0.5E_2$$

$$G_{23} = 0.2E_2 \quad \upsilon_{12} = 0.25$$

Dimensionless buckling load parameter:

$$\lambda_{\bar{N}} = \frac{\bar{N}a^2}{E_2 h^3} \tag{6.6}$$

6.3.3 New numerical examples

After validating the accuracy of the current numerical model, it has been expanded to explore the new numerical examples of the buckling behaviour of a sandwich panel. The analysis considers a square (l/b = 1) sandwich flat panel with a lay-up of (0°/45°/C/45°/0°), as shown in Figure 6.4. The panel has a thickness ratio of (l/b = 50), a core–face thickness ratio of (h_c/h_f = 16), is subjected to uniaxial axial compressive load (at x' = l), and is clamped at one end unless otherwise specified. The material properties of the sandwich structures used in further investigation are outlined in Table 6.2. The critical buckling load of the sandwich structures is presented in dimensionless form using Equation (6.6).

Table 6.2 The sandwich panel's material properties

Material properties	Face sheet	Core
$E_1 (Pa)$	25×10^9	0.04×10^9
$E_2 (Pa)$	1×10^9	0.04×10^9
$G_{12} (Pa)$	0.5×10^9	0.016×10^9
$G_{13} (Pa)$	0.5×10^9	0.06×10^9
$G_{23} (Pa)$	0.2×10^9	0.06×10^9
$\upsilon_{12} = \upsilon_{13}$	0.25	0.25
υ_{23}	0.25	0.25
$\rho \left(kg/m^3 \right)$	1,000	94.195

Source: Ref. [27].

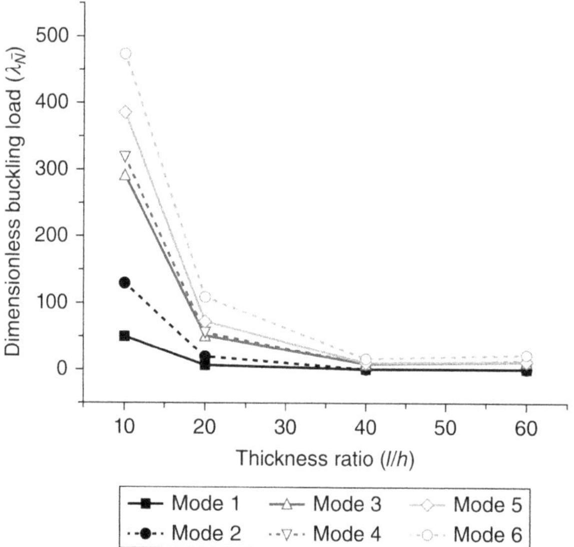

Figure 6.5 Influence of thickness ratio on the dynamic stability of a sandwich plate.

6.3.3.1 Influence of thickness ratio on the buckling behaviour of a sandwich flat panel

In this particular section, the variation of thickness ratios (l/h = 10, 20, 30, 40, 50, and 60) has been considered to examine the dynamic stability of a flat sandwich panel. The dimensionless critical buckling load has been computed and is plotted in Figure 6.5. From the plotted figure, it is evident that the critical buckling load decreases with an increase in the thickness ratio for all six modes. This phenomenon may occur because, as the thickness ratio of the panel increases, the overall stiffness of the plate decreases.

6.3.3.2 Influence of aspect ratio on the buckling behaviour of a sandwich flat panel

Within this section, an illustrative example has been attempted to examine the impact of the aspect ratio (l/b = 0.5, 1, 1.5, and 2) on the non-dimensional buckling characteristics of a flat sandwich structure. The outcomes have been calculated and are visually depicted in Figure 6.6. It is evident from the results that the buckling loads decrease with an increase in the aspect ratio, aligning with the predicted trend. This trend arises because, with an increased aspect ratio, the panel's geometry shifts from square to rectangular, with two sides of the panel becoming more distant from each other (with b remaining constant), consequently leading to a decrease in the overall stiffness of the panel.

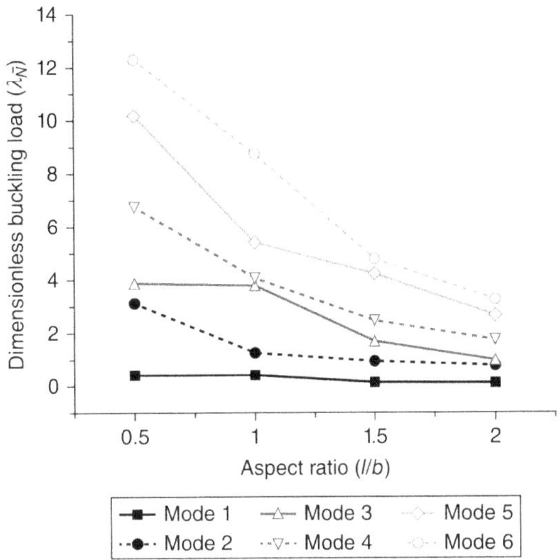

Figure 6.6 Influence of aspect ratio on the dynamic stability of a sandwich plate.

6.3.3.3 Influence of core to face sheet thickness ratio on the buckling behaviour of a sandwich flat panel

In this example, the influence of core–face thickness (h_c/h_f = 4, 8, 12, and 16) ratio on the dynamic stability of a sandwich panel has been studied. Figure 6.7 provides an excellent representation of the impact of the core–face thickness ratio on the sandwich panel's buckling behaviour. The critical buckling

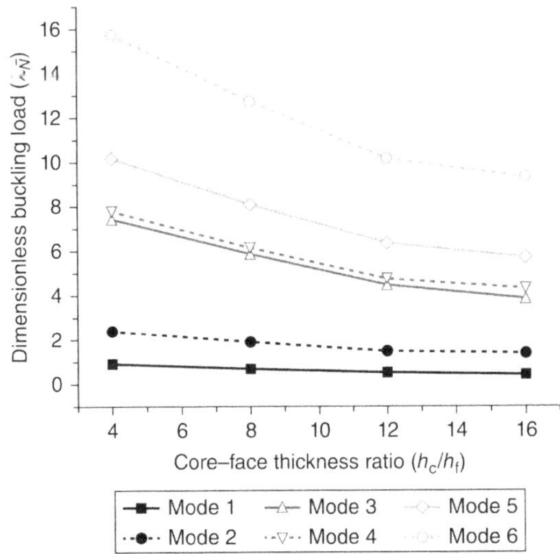

Figure 6.7 Influence of core-face thickness ratio on the dynamic stability of a sandwich plate.

load is clearly shown to decrease with a rise in the core–face thickness ratio. This decline in the critical buckling load can be attributed to the larger core–face thickness ratio, resulting in an overall reduction in the panel's stiffness. Essentially, as the core–face thickness ratio rises, the panel becomes less resistant to buckling, underscoring a critical consideration in the design and evaluation of the stability of such sandwich structures.

6.3.3.4 Influence of modular ratio on the buckling behaviour of a sandwich flat panel

This example illustrates the impact of the orthotropic ratio, for face sheet, on the dynamic stability of a sandwich composite plate. Eigenvalues were computed for five distinct modular ratios (E_1/E_2 = 15, 20, 25, 30, and 35) and are depicted in Figure 6.8. The figures distinctly reveal that the critical buckling loads rise as the modular ratios increase. In this specific case, Young's modulus in the transverse direction remains constant, while the longitudinal elastic modulus varies to achieve diverse modular ratios. Consequently, the longitudinal Young's modulus increases with the rise in modular ratios, leading to a subsequent increase in global stiffness values. This, in turn, amplifies the buckling loads across all six modes concurrently. Figure 6.9 presents all six mode shapes, illustrating the computed buckling behaviour of the composite panel with an orthotropic ratio of 15.

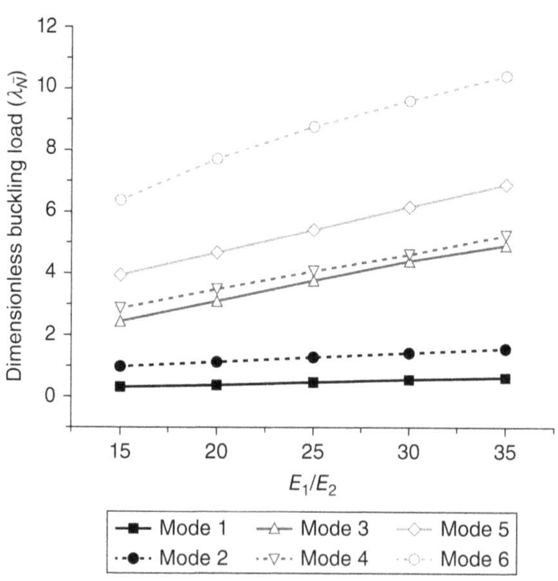

Figure 6.8 Influence of modular ratio on the dynamic stability of a sandwich plate.

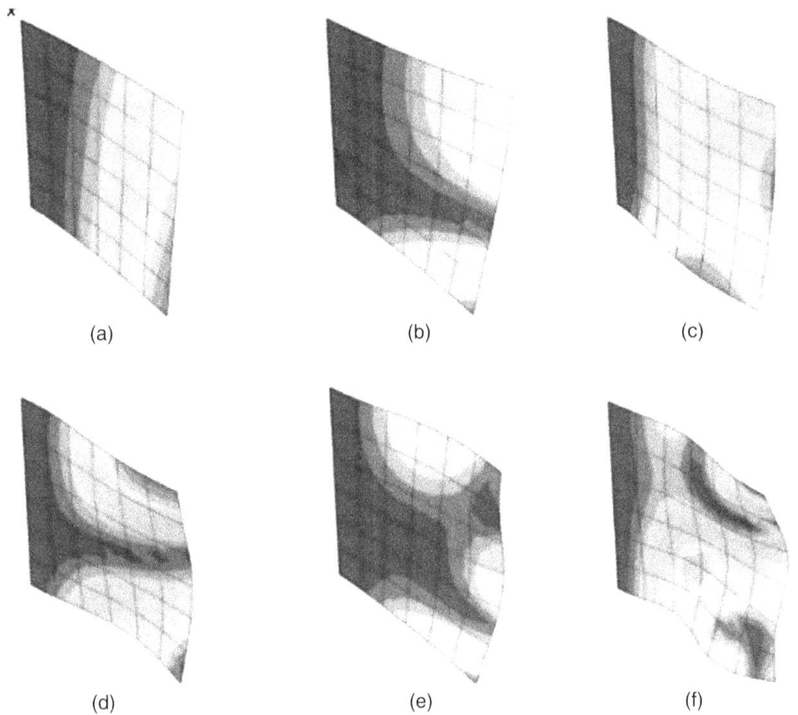

Figure 6.9 Mode shapes of sandwich composite: (a) 1st mode shape (b) 2nd mode shape (c) 3rd mode shape (d) 4th mode shape (e) 5th mode shape (f) 6th mode shape.

6.4 CONCLUSIONS

This comprehensive study focuses on the numerical analysis of the buckling behaviour of sandwich composite plates under uniaxial mechanical loads. The analysis relies on FSDT and FEM. The widely utilized finite element software, ABAQUS, is employed to execute this investigation. To ensure the validity of the numerical model, an initial verification process involves a meticulous comparison of computed results with available data from published literature. Following this validation phase, the study expands to explore the effects of various structural parameters on buckling responses through a series of new examples. The results obtained from these new cases are presented and discussed in the subsequent sections.

- The following outcomes offer insights into the dimensionless buckling behaviour of a flat sandwich structure, considering the influence of geometrical parameters as assessed via the FEM.
- The results emphasize the substantial impact of aspect ratio, thickness ratio, core–face thickness ratio, and end conditions on the critical buckling load.

- Also, the plate thickness ratio significantly influences the structure's buckling load. Lower thickness ratios yield the maximum critical load, while higher thickness ratios result in the minimum critical load.
- Furthermore, with an increase in the aspect ratio (while keeping b constant), the critical buckling loads decrease across all six modes.
- The dynamic stability is also affected by the core–face sheet thickness ratio. An increase in the core–face thickness ratio may result in a decrease in the critical buckling load.
- The higher orthotopic ratios (E_1/E_2) yield the maximum critical buckling load, whereas the lower orthotropic ratio corresponds to the minimum.

6.5 LIMITATIONS

- This research focuses on investigating the buckling behaviour of the sandwich plate structure under mechanical load (thermal and hygrothermal have not been considered).
- The analysis has only been performed for the infinite curved/flat panel structure.
- Furthermore, the stability analysis of sandwich flat panels has not taken into account the effect of cut-outs.

6.6 FUTURE RESEARCH DIRECTION

- Currently, researchers have delved into the buckling analysis of sandwich composite plates. Moreover, this analysis can be expanded to include bending and vibration analyses for both plate and shell panels.
- The impact of cut-outs on the buckling responses of the sandwich plate can be explored.
- The influence of hygrothermal loading on the static analysis of the sandwich panel can be studied.
- The influence of edge constraint conditions on the buckling behaviour of the sandwich plate can be explored.

REFERENCES

1. Afonso, A., Tomé, M., Moleiro, F., and Araújo, A. L., "Dynamic Buckling of Active Sandwich Panels," *Composite Structures*, Vol. 322, No. July, 2023, p. 117355. https://doi.org/10.1016/j.compstruct.2023.117355
2. Jeyakrishnan, P. R., Chockalingam, K. K. S. K., and Narayanasamy, R., "Studies on Buckling Behavior of Honeycomb Sandwich Panel," *The International Journal of Advanced Manufacturing Technology*, Vol. 65, Nos. 5–8, 2013, pp. 803–815. https://doi.org/10.1007/s00170-012-4218-9
3. di Sciuva, M., "Bending, Vibration and Buckling of Simply Supported Thick Multilayered Orthotropic Plates: An Evaluation of a New Displacement Model," *Journal of Sound and Vibration*, Vol. 105, No. 3, 1986, pp. 425–442. https://doi.org/10.1016/0022-460X(86)90169-0

4. Chai, G. B., and Khong, P. W., "The Effect of Varying the Support Conditions on the Buckling of Laminated Composite Plates," *Composite Structures*, Vol. 24, No. 2, 1993, pp. 99–106. https://doi.org/10.1016/0263-8223(93)90031-K

5. Khdeir, A. A., and Reddy, J. N., "Buckling of Cross-Ply Laminated Beams with Arbitrary Boundary Conditions," *Composite Structures*, Vol. 37, No. 1, 1997, pp. 1–3. https://doi.org/10.1016/S0263-8223(97)00048-2

6. Gilat, R., Williams, T., and Aboudi, J., "Buckling of Composite Plates by Global–Local Plate Theory," *Composites Part B: Engineering*, Vol. 32, No. 3, 2001, pp. 229–236. https://doi.org/10.1016/S1359-8368(00)00059-7

7. Singh, B. N., Iyengar, N. G. R., and Yadav, D., "Effects of Random Material Properties on Buckling of Composite Plates," *Journal of Engineering Mechanics*, Vol. 127, No. 9, 2001, pp. 873–879. https://doi.org/10.1061/(ASCE)0733-9399(2001)127:9(873)

8. Shukla, K. K., and Nath, Y., "Buckling of Laminated Composite Rectangular Plates under Transient Thermal Loading," *Journal of Applied Mechanics, Transactions ASME*, Vol. 69, No. 5, 2002, pp. 684–692. https://doi.org/10.1115/1.1485755

9. Ungbhakorn, V., and Singhatanadgid, P., "Buckling Analysis of Symmetrically Laminated Composite Plates by the Extended Kantorovich Method," *Composite Structures*, Vol. 73, No. 1, 2006, pp. 120–128. https://doi.org/10.1016/J.COMPSTRUCT.2005.02.007

10. Singh, S., Singh, J., and Shukla, K. K., "Buckling of Laminated Composite Plates Subjected to Mechanical and Thermal Loads Using Meshless Collocations," *Journal of Mechanical Science and Technology*, Vol. 27, No. 2, 2013, pp. 327–336. https://doi.org/10.1007/S12206-012-1249-Y

11. Neves, A. M. A., and Ferreira, A. J. M., "Free Vibrations and Buckling Analysis of Laminated Plates by Oscillatory Radial Basis Functions," *Curved and Layered Structures*, Vol. 3, No. 1, 2016, pp. 17–21.

12. Xing, Y., and Xiang, W., "Analytical Solution Methods for Eigenbuckling of Symmetric Cross-Ply Composite Laminates," *Chinese Journal of Aeronautics*, Vol. 30, No. 1, 2017, pp. 282–291. https://doi.org/10.1016/J.CJA.2016.12.027

13. Singh, D. B., and Singh, B. N., "New Higher Order Shear Deformation Theories for Free Vibration and Buckling Analysis of Laminated and Braided Composite Plates," *International Journal of Mechanical Sciences*, Vols. 131–132, 2017, pp. 265–277. https://doi.org/10.1016/J.IJMECSCI.2017.06.053

14. Nima, S. J., and Ganesan, R., "Buckling Analysis of Symmetrically Laminated Composite Plates Including the Effect of Variable Pre-Stress Field Using the Ritz Method," *European Journal of Mechanics: A/Solids*, Vol. 90, 2021, p. 104323. https://doi.org/10.1016/J.EUROMECHSOL.2021.104323

15. Higginson, K., Fernando, D., Veidt, M., Burnton, P., You, Z., and Heitzmann, M., "Local Buckling of FRP Thin-Walled Plates, Shells and Hollow Sections with Curved Edges and Arbitrary Lamination," *Thin-Walled Structures*, Vol. 168, 2021, p. 108242. https://doi.org/10.1016/J.TWS.2021.108242

16. Chen, W., Yang, J. S., Wei, D. Y., Yan, S. T., and Peng, L. X., "Buckling Analysis of Corrugated-Core Sandwich Plates Using a FSDT and a Meshfree Galerkin Method," *Thin-Walled Structures*, Vol. 180, No. September, 2022, p. 109846. https://doi.org/10.1016/j.tws.2022.109846

17. Ye, W., Liu, J., Zang, Q., Yin, Z., and Lin, G., "Buckling Analysis of Three-Dimensional Functionally Graded Sandwich Plates Using Two-Dimensional Scaled Boundary Finite Element Method," *Mechanics of Advanced Materials and Structures*, Vol. 29, No. 17, 2022, pp. 2468–2483. https://doi.org/10.1080/15376494.2020.1866125

18. Sarafraz, M., Seidi, H., Kakavand, F., and Viliani, N. S., "Free Vibration and Buckling Analyses of a Rectangular Sandwich Plate with an Auxetic Honeycomb Core and Laminated Three-Phase Polymer/GNP/Fiber Face Sheets," *Thin-Walled Structures*, Vol. 183, 2023, p. 110331. https://doi.org/10.1016/j.tws.2022.110331

19. Xinyi, L., Yifeng, Z., Rong, L., and Evrard, I. A., "Static and Global Buckling Analysis of Sandwich Panels with Improved Star-Shaped Honeycomb Using VAM-Based Downscaling Model," *Composite Structures*, Vol. 323, 2023, p. 117458. https://doi.org/10.1016/j.compstruct.2023.117458

20. Qian, H., He, Y., Alamri, S., Pan, X., Wang, X., Baghaei, S., and Rezaie, R., "Buckling Analysis of Sandwich Plates Reinforced with Various Loading of Multi-Walled Carbon Nanotubes," *Engineering Structures*, Vol. 294, 2023, p. 116765. https://doi.org/10.1016/j.engstruct.2023.116765

21. Bui, X.-B., Nguyen, T.-K., and Nguyen, P. T. T., "Stochastic Vibration and Buckling Analysis of Functionally Graded Sandwich Thin-Walled Beams," *Mechanics Based Design of Structures and Machines*, Vol. 52, No. 4, 2024, pp. 2017–2039. https://doi.org/10.1080/15397734.2023.2165101

22. Ebrahimi, F., Ahari, M. F., and Dabbagh, A., "Stability Analysis of a Sandwich Composite Magnetostrictive Nanoplate Coupled with FG Porous Facesheets," *Acta Mechanica*, Vol. 235, 2024, pp. 2575–2597. https://doi.org/10.1007/s00707-023-03837-3

23. Kumar, R., Tiwari, S., and Kumar Hirwani, C., "On Transient Responses of Sandwich Plate with Cutout Using FEM," *Materials Today: Proceedings*, 2023. https://doi.org/10.1016/j.matpr.2023.05.409

24. Putcha, N. S., and Reddy, J. N., "Stability and Natural Vibration Analysis of Laminated Plates by Using a Mixed Element Based on a Refined Plate Theory," *Journal of Sound and Vibration*, Vol. 104, No. 2, 1986, pp. 285–300. https://doi.org/10.1016/0022-460X(86)90269-5

25. Noor, A. K., and Burton, W. S., "Assessment of Shear Deformation Theories for Multilayered Composite Plates," *Applied Mechanics Reviews*, Vol. 42, No. 1, 1989, pp. 1–13. https://doi.org/10.1115/1.3152418

26. Ferreira, A. J. M., Roque, C. M. C., Neves, A. M. A., Jorge, R. M. N., Soares, C. M. M., and Reddy, J. N., "Buckling Analysis of Isotropic and Laminated Plates by Radial Basis Functions According to a Higher-Order Shear Deformation Theory," *Thin-Walled Structures*, Vol. 49, No. 7, 2011, pp. 804–811. https://doi.org/10.1016/J.TWS.2011.02.005

27. Kumar, R., and Hirwani, C. K., "Investigation on Nonlinear Bending Behavior of Sandwich Shell Panel with Cutout Using HSDT and FE Approach," *Mechanics Based Design of Structures and Machines*, Vol. 52, No. 3, 2022, pp. 1–19. https://doi.org/10.1080/15397734.2022.2158868

Chapter 7

Recent trends in coconut coir fibre-reinforced composite material

Rahul Kumar, Faladrum Sharma, Sumit Bhowmik, and P. M. G. Bashir Asdaque

7.1 INTRODUCTION

To promote enhanced coordination on environmental issues and to create policies and programmes that drive sustainable development and improve human well-being, UN environment programme has launched a Green Economy Initiative (GEI). In the green economy, economic growth is driven by reduced carbon footprints, enhanced energy efficiency, and protected biodiversity and ecosystem [1–2]. This has enforced the governments and private organizations to reform policies and provide much capital in the research and development of greener products made of natural renewable resources. Being one of the most important classes of engineering materials, the remarkable shift is observed in the reinforcing phase of polymer composites from synthetic fibre to natural cellulosic fibres to fabricate eco-friendly composite material [3–5]. The polymer composite materials based on natural plant-based fibres depicted distinctive advantages concerning ecology along with economic perspectives. The principal sources of natural plant-based fibres are agriculture by-products and crops obtained after their primary processing [6–7]. Examples of different natural fibres utilized in polymer composites are wood, sisal, hemp, bamboo, coir, flax, pineapple leaf, kenaf, jute, and luffa cylindrica [8–10]. Among the aforementioned lignocellulosic fibres, coir has shown a great potential for application in the composite structure. Moreover, there have been deliberate investigations by various researchers on coir fibre-reinforced polymer composites in substituting wood species-based plastic components or composite panels [11–12].

The fruit of the coconut tree is the primary source of coir fibre production from its outer husk. The coconut tree is a group of perennial climbers in the botanical family of Arecaceae, subfamily Arecoideae, tribe Cocoeae, genus *Cocos*, and species *Cocos nucifera* L. There are mainly two varieties of coconut cultivated throughout the world: tall variety and dwarf variety [13]. The average height and lifespan of tall variety are 15–18 m and 80–90 years, respectively. The average height and lifespan for dwarf variety are 5–7 m and 50–70 years, respectively. India is the third largest coconut producer in the world, cultivating approximately 2 billion coconuts from a

DOI: 10.1201/9781003564355-7

1.98-million hectare area in 2015–2016 [14]. Coconut fibre (CF) has been extensively used in making floor mats, doormats, brushes, mattresses, and geotextiles to prevent soil erosion [15–16]. Despite the aforementioned production volume attributes and technological advancement, there are quite a few unresolved complications that hamper the wide-scale utilization of coir fibres:

- Insufficient and deficient consolidated literature background concerning the physical, chemical, and mechanical properties of coir fibre at the fundamental and molecular level.
- Missing general methodology for preparation and processing of macro-, micro-, and nano-fibre and filler from coconut husk and shells that can be as scientifically sound as well as commercially viable.
- Inconsistent interface functionalities and properties due to weak interphase interaction.

There are some research articles published earlier in this domain [17–19], but very few authors have concentrated on the aforementioned concerns. This chapter attempts to provide a consolidated overview on the properties and structure of coconut coir fibres. The importance of interface interaction between fibre and matrix that improves the performance of composite material has been highlighted. Furthermore, some approaches are suggested for surface functionalization of coir fibres to expand the existing research interests and commercialization of coir-based composite products. Also, an approach has been taken to highlight the importance of the biodegradable composite and bio-based polymer composite using coir filler and fibre as reinforcement. The role of the different types of matrix and variation in the matrix percentage of thermosetting, thermoplastic, and elastomers is discussed to understand the physical and mechanical behaviour of the developed composite materials.

7.2 STRUCTURE OF MULTI-SCALE COCONUT COIR FIBRES

The following sections describe the composition, structure and morphology coir fibres. Different types of extraction procedure and their working mechanism are highlighted.

7.2.1 Morphology and structural composition of coir cell wall

The distribution of coir fibre cell walls comprise four principal layers. These layers are designated as primary wall and secondary wall. There are three secondary walls (S_1, S_2, and S_3) and one outer primary wall (P), as shown

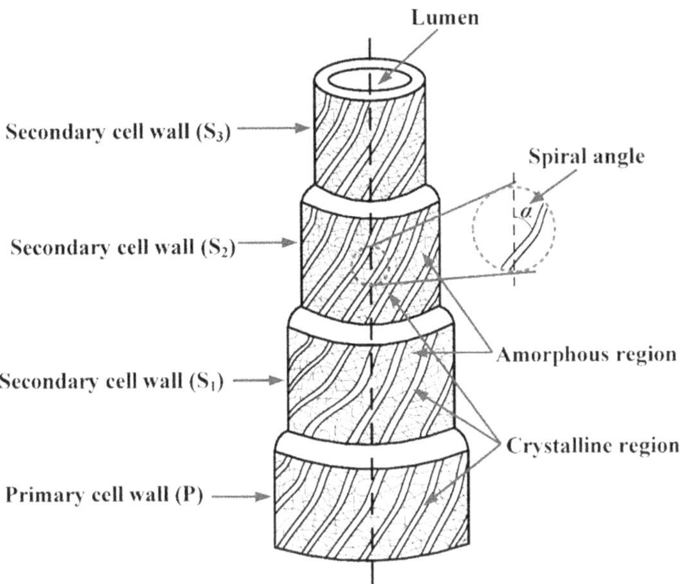

Figure 7.1 Coir fibre structure with different layers. (Modified from Reference [105] under the terms of Creative Commons Attribution License.)

in Figure 7.1. The three secondary walls are inner (S_3), middle (S_2), and outer (S_1) [20]. The outer primary wall (P) is extremely thinner and comparatively featureless with circular pit openings. The outer and middle secondary walls are thicker and have a similar thickness. The approximate thickness of the S_1 and S_2 layers lies between 1.1 and 1.2 µm. The outer and middle secondary layers contain a well-defined fibrillar pattern in which two different helical structures are formed by microfibrils. The S-helix and Z-helix are respectively formed in the outer (S_1) and middle (S_2) secondary walls. The inner secondary wall (S_3) is thinner (thickness ~ 0.089 µm) and does not reveal a definite fibrillar pattern like in the case of S_1 and S_2. However, some pit openings in the form of slit-shaped or circular shapes appeared on S_3. These pit openings were found to be oriented with a longitudinal fibre axis at 48°. It is also referred to as spiral angle represented by α in Figure 7.1. The overall macrostructure of coir fibres comprises a hollow tube with micrometre-sized fibre bundles organized in primary and secondary layers and individual fibres inserted in the lignin matrix.

7.2.2 Extraction and processing of coir fibres

The coconut coir fibres can be developed into various feed stock materials for commercial product development using various extraction and processing methodologies. The coconut palm tree might be considered as a vital fibre producing resource due to the fibre extraction from many parts of the tree like leaf sheath, midribs of leaves, or outer husk of the coconut fruit. Mainly two types of coir fibres are obtained from the coconut fruit nutshell: brown fibres from full-fledged coconuts and white fibres from underdeveloped coconuts. While white fibres have a finer and smoother surface, brown fibres are highly abrasion resistant and stronger compared to white fibres. For industrial applications, among long (bristle), short (mattress), and mixed fibres availability, brown fibres are highly utilized. The fibre produced from the husk of coconut fruit is the main source of industrial raw material. The coir fibres are extracted from the outer husk of the coconut fruit through a mechanical or retting process.

In mechanical extraction process, fibres are extracted from the outer husks using either defibring or decorticating equipment, as shown in Figure 7.2, developed by researchers at Central Coir Board, India. Initially, husks are immersed in water tanks for five days and then the husks are crushed in a beaker to open the fibre nodes. Afterwards, the revolving drum separates coarse lengthy fibres from the coir pith and short woody parts. The long fibres are then cleaned, washed, dried, hackled, combed, and cut into the required length for reinforcement [106].

Another process is retting and traditional extraction process. This process is time-consuming as well as painstaking, in which coconut husks are left for retting for the duration of three to six months in lagoons or brackish water. The retting partially decomposes the pulp of the shell. The anaerobic

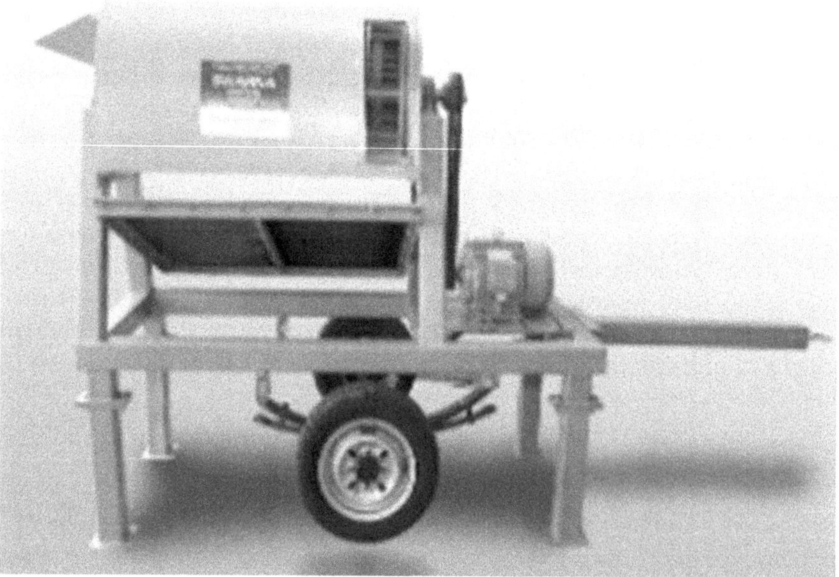

Figure 7.2 Coir fibre extraction machine [103].

fermentation (for 10–12 months) retted the husks and softened husks are beaten by a hammer, ripped from the husks and separated with a comb to extract fibres. The extracted fibres are then hackled, cleaned, washed, and dried in shade. This kind of fibre extraction practice results in a high-quality yield of white coir fibres for spinning and weaving. The shorter periods of retting are applied for the production of brown fibres, used in geo-textiles applications.

7.2.3 Structure and properties of coir fibres

7.2.3.1 Chemical composition and crystalline structure

Similar to other plant-based bio-fibres, major components of coir fibres are hemicellulose, cellulose, and lignin. The average composition of these important constituents varies from 32% to 50% for cellulose, 30–46% in the case of lignin, and 0.15–15% for hemicellulose. Other variations can be in form of pectin with 3–4%. The chemical composition variation for coir fibres is documented by different researchers and presented in Table 7.1. It can be observed from the constituent % variation that the lignin matrix along with cellulose fibres completes 75–95% of the composition [21–25]. Lignin is a major contributor towards increasing the thermal stability of fibres, whereas cellulose provides the mechanical strength/toughness. Cellulose is polysaccharide macromolecules, consisting of a variable number of $\beta(1\rightarrow4)$ linked D-glucose units, whereas lignin is a phenolic-based cross-linked polymer. Moreover, the sources of origin, location of trees/plants, and extraction methods affect the composition of coir fibres.

The crystal structure of any material can be determined using X-ray diffraction analysis [26]. The cellulosic part of coir fibres demonstrates crystalline behaviour, whereas amorphous nature is observed in the lignin section. Two crystalline peaks can be perceived at $2\theta = 10°, 22°$. Moreover, the crystallinity % of coir fibres usually varies from 60% to 78% with the lowest for the raw fibres. Also, with the employment of various types of chemical treatment, the crystallinity index and % got significantly improved owing to the frequent removal of lignin from molecular structure; therefore, more ordered arrangement in crystals comes up. There are two types of coir fibres such as white fibre, a novice form, and brown fibre, a more matured form. Mature brown coir fibres contain more lignin and less cellulose than white fibres and

Table 7.1 Composition of coir fibres reported in previous literatures

Cellulose (%)	Hemicellulose (%)	Lignin (%)	Pectin (%)	Reference
36–43	0.2	41–45	1.8	[22]
42.14	15–17	35.25	–	[94]
32–43	0.15–0.25	40–45	–	[96]
38–46	10–15	37–41	–	[21]

Table 7.2 Mechanical properties of coir fibres reported in previous literatures

Tensile strength (MPa)	Young's modulus (GPa)	Elongation (%)	Reference
131–175	4–6	15–40	[11]
95–230	2.2–6	15–51.4	[99]
59–297	3.2–4.3	–	[100]
144.6	–	32.3 ± 0.2	[98]

so are stronger but less flexible. They are made up of small threads, each less than 0.05 inch (1.3 mm) long and 10–20 μm in diameter [27].

7.2.3.2 Mechanical and physical properties of coir fibres

The mechanical properties like ultimate tensile strength, elastic modulus, strain at break impact resistance, and fracture toughness of coir fibres are presented in Table 7.2. It can be observed from the demonstrated data that the tensile strength of coir fibres varies from 131 MPa to 175 MPa as reported by different researchers. Therefore, the average value of tensile strength is 152 MPa; of course, the origination of fibre is one of the prime factors in strength determination. On the other hand, Young's modulus values showed variation from 4 GPa to 6 GPa with a mean value of 5 GPa. These values depict the comparability of coir fibre with respect to their synthetic counterpart like glass in terms of strength and stiffness. Moreover, strain at break values depicted variation from 15% to 51%, as seen in Table 7.2.

The physical properties like density, appearance, textures, length, and diameter of coir fibres are evaluated by different researchers. All these properties are depicted in Table 7.3. The density of coir fibres varies from 1.15 g/cm³ to 1.5 g/cm³ with an average value equal to 1.322 g/cm³. The typical diameter range of coir fibres varies from 0.01 mm to 0.46 mm. Moreover, in other work, the reported diameter is 10–460 μm. The length is 718.8 μm. In another work, the length varies from 20–150 mm. In some work, diameter varies from 200 ± 10 μm. The density is 1.2 ± 0.1 g/cm³. The moisture gain is around 10% at a relative humidity of 65%. The appearance and textures of coir fibres depend on the method of extraction. However, both smooth and rough textures can be found on the brown fibres.

Table 7.3 Physical properties of coir fibres reported in previous literatures

Average diameter (mm)	Water absorption (%)	Density (g/cm³)	Reference
0.01–0.46	–	1.15–1.46	[22]
0.1–0.45	10	1.3–1.5	[94]
0.38	–	1.2	[96]
0.1–0.4	–	1.15	[97]
0.2 ± 0.01	–	1.2 ± 0.1	[98]

7.3 COCONUT FIBRE (CF)-REINFORCED POLYMER COMPOSITES

The coir fibres are consumed as a reinforcing material in the composite development owing to being extremely resistant to harsh weather conditions. Additionally, abundant availability as agricultural by-products and relatively low cost compelled the researchers and material scientists to focus their attention on coir fibres and explore their industrial and commercial employability. In the last decade, several research works have been carried out in this field and the different factors that are influencing the various properties of composites have been studied. Elaborative studies were carried out to assess the background information on the issues to be considered in this work and to emphasize the relevance of the present study. The polymers utilized most often for coir fibre-based composite production are comprised of biodegradable plastics, thermoset plastic materials, thermoplastics, and rubbers. In the construction industry, coir fibres are being reinforced with cement to improve the tensile properties of the reinforcing concrete.

7.3.1 Biodegradable matrix-based composite

The thermoplastic polymer matrices like thermoplastic starch, polylactic acid (PLA), poly(butylene succinate) (PBS), wheat gluten, and polycaprolactone (PCL) are originated from biomass. These biodegradable polymer matrices are established as a potential substitute for petroleum-based plastics. When natural fibres are mixed and reinforced in the biodegradable matrix, it produces bio-composites that are considered fully environment-friendly owing to the full decomposing of all its parts into CO_2 and H_2O. Recently, coconut fruit-based coir fibres have attentively emerged as an alternative to other natural plant fibres for the development of bio-plastics.

Iovino et al. prepared thermoplastic starch and short coir fibre-based composite and investigated its aerobic biodegradation under controlled composting conditions. The aerobic biodegradation of starch and polylactic acid (75 wt.%)–starch (25 wt.%) matrix was compared and results showed the starch being more bio-susceptible material as fully biodegraded by the completion of the incubation process [6]. Jang et al. analysed the thermophysical and mechanical properties of plasma-treated CF-reinforced poly(lactic acid) composite and reported substantial improvement in mechanical properties. There is a considerable enhancement in interfacial adhesion of coir fibres after plasma treatment is done [7]. Dong et al. characterized the mechanical, thermal, and biodegradable properties of PLA-coir fibre biocomposites as a function of fibre content (5–30 wt.%). Additionally, fibres were treated chemically to improve their surface morphology to achieve good adhesion between matrix and fibre. With the increase in fibre content, tensile and flexural strengths were observed to be decreased with 20 wt.% fibre content emerging as optimum value [8]. Diao prepared another biocomposite by using toughened wheat gluten as a matrix with reinforcement of surface-modified coir fibre. The fabricated composite showed promising mechanical properties with enhanced work to failure value (from 1.45 MJ/m^3 to 4.93 MJ/m^3). With 15 wt.% of fibre content, flexural stress value increased from 46 MPa in neat wheat gluten to 71 MPa for composite samples [12]. Hemsri et al. reported an enhancement of 80% in tensile strength of wheat gluten composite with reinforcement of chemically modified CFs. The chemical modification was done initially with alkali treatment, subsequently followed by silane treatment. This hybridization or dual surface modification was observed to be quite effective in enriched interfacial adhesion between fibre and matrix [13]. Corradini et al. fabricated a hybrid biodegradable polymer matrix composite using reinforcement of jute, sisal, and coconut along with starch and gluten matrix. The fibre reinforcement weight was varied from 5% to 30% of the total polymer. They reported a decrease in the moisture/water absorption and an increase in tensile strength/ Young's modulus value of matrix after the inclusion of lignocellulosic fibres [14]. Ramírez et al. characterized the green coconut coir fibres-reinforced starch composite with and without surface treatment. It was found that the tensile properties of the starch matrix got enhanced with fibre inclusion and surface modification [15]. Nam et al. prepared surface-modified coir fibre-reinforced polybutylene succinate (PBS-coir) biocomposites. The coir fibres were incorporated with mass content of 10%, 15%, 20%, 25%, and 30%. The surface modification of fibres was done with a 5% NaOH solution. The results showed the best mechanical properties at 25% of fibre content [16]. Chun et al. prepared the polylactic acid composite reinforced with coconut shell powder and investigated the influence of filler content and acrylic acid on mechanical and thermal properties of biocomposites. The acrylic acid was incorporated as a chemical modifier for the betterment of adhesion between filler and matrix. The tensile strength and elongation at break decreased with the inclusion of coconut shell powder. However, Young's modulus value was improved. Moreover, the treatment of coconut shell powder with acrylic acid resulted in enhanced tensile strength and elastic modulus along with reduced elongation value [29].

7.3.1.1 Thermosetting matrices composite

The thermosetting matrices used most often for the development of coir fibre/ filter-based composite are epoxy, phenolic resin, and unsaturated polyester. The research works carried out on these matrix materials are deliberated in the following sections.

7.3.1.1.1 Epoxy-based composite

Harish et al. developed and evaluated the mechanical properties of coir fibre-reinforced epoxy composites. The average values of tensile, flexural, and impact strength of developed composites were 17.86 MPa, 31.08 MPa, and

11.49 kJ/m^2, respectively [30]. Biswas et al. also fabricated coconut coir-based epoxy composite to investigate hardness and other mechanical properties as a function of fibre length. It is found that the hardness value decreased with a rise in the fibre length up to 20 mm. The prepared composite material showed ultimate tensile strength, Young's modulus, flexural strength, and impact strength of 13.05 MPa, 2.064 GPa, 35.42 MPa, and 17.5 kJ/m^2, respectively [31]. Sarki et al. used coconut shell particle fillers to reinforce epoxy matrix and assessed the prospect of utilizing it as a novel material to develop polymer composites for various engineering applications. The composite was prepared with shell particle filters of up to 30 wt.%. The elastic modulus and ultimate tensile strength values were increased, but impact strength value dropped to some extent with coconut shell particle content [32]. Aireddy and Mishra experimentally investigated the abrasive and erosive wear behaviour of epoxy polymer composite reinforced with coconut coir dust. Erosive and abrasive wear rates decreased and increased respectively with an increase in coir dust amount. Moreover, with an increase in normal load, abrasive wear resistance drops [33].

Saw et al. prepared an epoxy-based hybrid composite with jute fibres (cellulose-rich, >65%) and coconut coir fibres (lignin-rich, >40%) and compared their mechanical properties with and without the chemical treatment of fibres. The highest value of tensile and flexural strength of the hybrid composite was observed for the jute–coir content of 50:50. Moreover, these values were 78% and 61% higher than those for untreated fibre composite. Therefore, the surface modification significantly impacted the composite properties [34]. Bhagat et al. investigated the physical, mechanical, and water absorption behaviour of coir-glass fibre-reinforced hybrid epoxy composite. With fibre length of 15 mm at 10 wt.% fibre content, the maximum value of flexural strength (63 MPa) was observed [35]. An epoxy polymer-based hybrid composite was prepared with coir pith and nylon fabric in different proportions. The hand lay-up technique along with compression moulding was used to fabricate the sample. Different mechanical properties like tensile, flexural, impact, and hardness of composite were assessed to investigate the effect of chemical treatment of coir pith and inclusion of nylon fabric. The mechanical properties were observed to be enhanced, thus confirming the positive impact of chemical treatment and nylon fabric reinforcement [36]. In another work, walnut shell particles (20 wt.%) along with coconut coir fibres (10 wt.%) were used to fabricate epoxy-based hybrid polymer composite. The composite material was fabricated using the squeeze casting method and its mixed-mode fracture characteristics were investigated. The average fracture toughness values for modes I and II were 1.319 MPa·m$^{0.5}$ and 1.219 MPa·m$^{0.5}$, respectively [37].

The untreated and treated coconut sheath fibre-reinforced epoxy composites were prepared by hand lay-up process along with compression moulding technique. From the experimental results, it was found that the treated sheath fibre composites possess superior mechanical property and thermal stability as compared to untreated sheath fibre. There was an enhancement of 21%, 18%, and 24% in tensile, flexural, and impact strengths, respectively, for treated sheath fibre composite in comparison to untreated one [38]. Ojha et al. successfully fabricated the wood apple-based polymer composites and compared their mechanical and tribological properties with coconut shell particulate filler-reinforced composite. The results showed the respective maximum values of flexural strength for wood apple shell and coconut shell filler composites are 78.19 MPa and 68.25 MPa, respectively [39]. Prasad et al. studied and applied artificial neural network to predict the mechanical properties of the coir–epoxy composite. The fibre length is observed to be having a significant effect among various parameters on the mechanical properties [40]. Recently, Sharma et al. [104] fabricated epoxy composite samples reinforced with coir fillers in different weight percentages. Both untreated and alkali treated coir fillers were employed for preparing the samples. The samples were subjected to fatigue testing at different loading levels, namely 30%, 45%, and 60% of the ultimate strength of the fabricated samples. The fatigue life was the highest at 30% loading level for 10 wt.% untreated coir filler.

7.3.1.1.2 Unsaturated polyester-based composite

Monteiro et al. appraised the mechanical properties and structural characteristics of coir fibre-reinforced polyester composites. Coir fibres up to 80 wt.% were used to fabricate the composite. The composite samples act like toughened ones if reinforced up to 50 wt.% of fibre content. Further addition of coir fibres beyond aforesaid amount leads to the formation of agglomerates in samples and thus composites perform like more flexible one [41]. Rout et al. prepared polyester and propylene matrices-based coir composites using short as well as continuous coir fibres. Both untreated and treated composite specimens are fabricated. Different treatment methodologies like defatting, polymethyl methacrylate (PMMA) grafting, polyacrylonitrile (PAN) grafting, and cyanoethylation are utilized to modify the coir fibre surface. The hand lay-up and compression moulding techniques are used to prepare coir-polyester and coir-propylene composite samples, respectively. The tensile and impact properties are investigated for both types of composite samples [42]. The same group of researchers utilized the alkali treatment, bleaching, and vinyl grafting for the coir fibres modification. The impact, flexural, and tensile properties are found to be increased with fibre surface modification. Among the various modification methods, bleached fibre-reinforced polyester composite showed superior flexural strength (61.6 MPa). However, alkali treatment with 2 wt.% resulted in substantial enhancement in tensile strength (26.80 MPa) [43]. Prasad et al. used alkali treatment to enhance the wettability of coir fibres to the polyester matrix. The treated and untreated fibres in the polyester matrix developed the porosity of 5% and 11%, respectively, thus confirming the improved adhesion with the matrix. Moreover, there was an improvement of 15% in the tensile strength of fibre when treated with 5 wt.%

alkaline solution (NaOH) [44]. Hill and Khalil manufactured non-woven coir fibre-reinforced polyester composite with unmodified and modified fibres. The chemical modification was carried out by acetylation and silane treatment. The composite samples with treated and untreated fibres were subjected to a fungal decay test for up to 12 months. It was observed that the silane treatment showed a significant amount of resistance from decay [45]. In another research work, Hill and Khalil addressed the issue of compatibility between the plant fibres and polyester matrices [46]. Mulinari et al. studied the fatigue behaviour of alkali-treated coconut coir fibres-reinforced polyester composite. The coir fibres were surface-modified with 1% wt./v NaOH solution. The compression moulding method was used to fabricate composite samples. The experimental results showed a drop in the fatigue life of composite when subjected to higher tension [47]. Jayabal et al. prepared calcium carbonate particle-impregnated coir fibre-reinforced polyester composite by following design of experiments methodology. The fibre length, fibre diameter, and fibre content were chosen as three-level parameters. Moreover, a model using an artificial neural network was established to envisage the mechanical properties of composite [48]. Jayabal et al. investigated the mechanical properties of untreated and treated woven coir fibre-reinforced polyester composites. The average value of flexural and tensile strength for the untreated composite samples was 31.3 MPa and 19.9 MPa, respectively, whereas the impact strength is 49.9 kJ/m^2. However, for treated composite, there was substantial improvement of about 40%, 42%, and 20% in tensile, flexural, and impact strength values [49]. Jayabal and Natarajan evaluated and developed the mathematical model for flexural, tensile, and impact properties of coir fibre polyester composites. The fibre content had significant effect on various mechanical strengths than fibre length [50].

Yousif evaluated the frictional and wear performance of multilayer with three and four layers of coir fibre-reinforced polyester composites under dry sliding conditions. The input parameters for the tests are sliding distance and applied loads, whereas output parameters are specific wear rate and friction coefficient. The three-layer composite showed a better wear rate, whereas the four-layer composite revealed a greater coefficient of friction [51]. Owolabi et al. prepared the coconut hair fibre-reinforced phenol-formaldehyde and unsaturated polyester composite. The unsaturated polyester is applied as binder material to prepare bulk moulding compound type press material using coir fibres [52]. Rout et al. fabricated coir–polyester composite samples by varying fibre loading and studied their various mechanical strengths. The fibre surface was subjected to the dewaxing process by alkali treatment (5%). Moreover, methyl methacrylate grafting was carried out on alkali-treated fibres. The improvement of around 30% and 27% was observed in tensile and flexural strength of composite samples as compared to neat polyester samples [53]. Abdullah and Ahmad used treated CFs to prepare polyester composites. The treatment of fibres was carried out using alkali, silane, and silane on alkalized fibre. A substantial improvement in tensile properties of composite samples was observed after treatment [54]. Jappes et al. prepared composite laminates from naturally woven coconut sheath and compared their various mechanical properties with glass fibre composite. It was observed that coconut sheath-reinforced polyester composite performs superiorly in impact and flexural tests [55].

7.3.1.2 Thermoplastic matrices composite

Various thermoplastic matrices like polyethylene, polyvinyl chloride, polypropylene, and nylon are used to develop coir fibre-reinforced thermoplastic matrix composites. The previous research work done on these composites is elaborated in the following sections.

7.3.1.2.1 Polyethylene-based composites

The CF-reinforced low-density polyethylene composite was prepared and its interfacial bonding with matrix material was investigated by a single-fibre pull-out test. It was observed that the waxy layer present on the fibre surface acted as bonding force between fibre and matrix [56]. Arrakhiz et al. prepared sodium hydroxide (NaOH) and bromododecane (CBr)-treated coir fibre-reinforced high-density polyethylene (HDPE) composite. The comparison of mechanical properties of composite samples with neat polymer was carried out. A substantial improvement in torsional modulus and tensile properties was observed as compared to neat polymer. The elastic modulus value was enhanced by 120% [57]. In other research work, coconut flour-reinforced polyethylene composite was prepared by using surface-modified coconut flour along with various coupling agents. The NaOH wt.% used for flour surface modification were 5% and 18%. The treatment leads to superior tensile properties of the composite samples [58]. Fernandes et al. fabricated hybrid composite with cork powder and coconut short fibres as reinforcement and high-density polyethylene as matrix phase. A maleic anhydride-based coupling agent was used to improve adhesion between the reinforcing and matrix phase. The enhancement of about 27% and 47% was observed in elastic modulus and tensile strength values, respectively, after the inclusion of CFs in comparison with unreinforced composites [59].

7.3.1.2.2 Polyvinyl chloride-based composites

Leblanc investigated the rheological characteristics as well as hardness and impact resistance of CF-reinforced polyvinyl chloride composite. The prepared composite samples showed a non-linear viscoelastic character in the molten state as compared to virgin polyvinyl chloride's linear behaviour. The complex modulus value increased with the addition of green CF and there was no deterioration in mechanical properties of the composite up to fibre content of 30 wt.% [51–60]. Tran et al. implemented a micromechanical with a physical-chemical approach to examine interfacial compatibility of coir fibres with polyvinyl chloride and polypropylene matrices. It was observed that superior adhesion between coir fibres and polyvinyl chloride matrix was found as compared to polypropylene matrix due to increased physical–chemical interaction. Moreover, the flexural strength was also found to be higher in the longitudinal direction of coir fibre reinforcement in the composite [62].

7.3.1.2.3 Polypropylene-based composites

Leblanc et al. prepared coconut coir fibre-reinforced polypropylene composite and investigated its mechanical properties. The composite samples were fabricated in molten and solid states by utilizing various important techniques [63]. Rozman et al. made a hybrid composite by reinforcing polypropylene matrix with coconut and glass fibres. The experimental tests showed the reduction in tensile, flexural, and impact strengths of composite in contrast to neat polypropylene matrix. However, the elastic and flexural modulus values got enhanced with the incorporation of fibres in the matrix material. The irregularity in fibre size and incompatibility of fibres with matrix was the prime reason behind strength reduction [64]. Sharma et al. used various wt.% like 0, 15, and 30 of coir and glass fibres to prepare polypropylene-based polymer composites. The fabricated composite samples were tested for weathering properties under moisture conditions as well as ultraviolet exposure. The glass fibre-reinforced polypropylene samples showed higher resistance to moisture absorption and thickness swelling as compared to coir fibre-reinforced polypropylene (PP) samples [65].

Lai et al. dealt with the use of various treatment methods like alkali, stearic acid, potassium permanganate, and acetone to modify the coir fibre surface and later fabricate the polypropylene composite sample by reinforcing with treated coir fibres. It was observed that the modulus of elasticity and flexural modulus exhibited higher magnitude in the case of treated fibre reinforcement in contrast with untreated one [66]. In another comparison work, polypropylene-based polymer composites were developed by using palm and coir fibres separately for various fibre loadings. Before the reinforcement, benzene diazonium salt was used to treat the fibre surface for enhancing the interfacial adhesion with the matrix. The coir fibre composite samples were found to have superior mechanical properties as compared to palm fibres. The fibre reinforcement of 30 wt.% was observed to be showing better properties [67]. Gu utilized the alkali-treated brown coir fibres as reinforcement in polypropylene matrix to fabricate composite with a heat press. The alkali treatment was carried out with varying concentrations of sodium hydroxide (from 2% to 10%). The tensile strength of composite samples was observed to be decreased with an increased amount of alkali treatment [68]. Islam et al. fabricated coir fibre-reinforced polypropylene composite and assessed its physical–mechanical properties. The composite samples were fabricated for raw and *o*-hydroxybenzene diazonium salt-treated coir fibres. The treated coir fibre composite exhibited better mechanical properties in contrast with untreated fibres. Moreover, the increasing amount of filler content resulted in a drop in tensile strength for both untreated and treated composites [69]. Ayrilmis et al. explained the reason behind the significant improvement in the dimensional stability of coir fibre-reinforced polypropylene composite. The stated reason was the presence of a lignin layer on the coir fibre. The best composite properties were observed for the coir fibre content of 60 wt.% along with 37 wt.% and 3 wt.% of polypropylene matrix and MAPP, respectively. Any further addition of fibre amount leads to a drop in tensile and flexural strength of composite [70].

Bettini et al. assessed the influence of maleic anhydride-grafted polypropylene on the compatibility of coir fibres with polypropylene matrix. Furthermore, various wt.% (2, 4, 8, and 12) were used for the chemical pulping of coir fibres. They also calculated the lignin concentration in coir fibre before and after the pulping. It was found that the tensile strength got enhanced significantly with 2 wt.% of NaOH concentration and afterwards the fall in the properties was observed [71]. Morandim-Giannetti et al. prepared the CF-reinforced polypropylene composite and examined its mechanical, thermal, and morphological properties. The properties were investigated with and without the presence of a compatibilizer along with lignin incorporation. In the presence of maleic anhydride-grafted propylene compatibilizer, the lignin inclusion resulted in the reduction in tensile strength of composite [72]. Haque et al. fabricated coir fibre-reinforced PP composite and investigated the dispersion behaviour of fibre in the matrix. The fibres were subjected to chemical treatment before reinforcement. The chemical treatment lead to minimized agglomerates formation, thus improving dispersion of the fibre in the polypropylene matrix. Also, the water uptake capacity of composite got reduced after treatment owing to the decline in hydroxyl groups in fibre surfaces [73]. Arrakhiz et al. examined the effect of alkali-treated coir, bagasse, and alfa fibre reinforcement on mechanical properties of polypropylene composite. The use of coir, alfa, and bagasse fibres lead to a 56–75% increase in elastic modulus as compared to virgin polypropylene matrix. On the other hand, 30–47% improvement in flexural modulus was observed with various fibre reinforcement [74]. Nandi et al. used coir fibres along with an *m*-isopropenyl-based coupling agent for the fabrication of polypropylene composite material. The concentration of the coupling agent was varied during the sample preparation and its effect on the tensile, impact, and flexural properties was investigated. All three mechanical strengths were found to be increased with coupling agent amount with 5% as optimum value [75]. Mir et al. fabricated polypropylene composites using both raw and treated coir fibres with a fibre content of 10 wt.%, 15 wt.%, and 20 wt.%. The treatment of coir fibres was carried out in an acidic medium with chromium sulphate and sodium bicarbonate salt. The tensile strength for both types of samples initially improved and then dropped with further addition in fibre content beyond 15%. Also, the treated coir fibre composite sample showed superior mechanical properties [76]. Zaman et al. prepared polypropylene composite with reinforcement of irradiated jute and irradiated coir fibres. The jute-based composite samples exhibited enriched mechanical properties in contrast with coir-based composites [77].

Bettini et al. assessed the fatigue and fracture behaviour of coir fibre-reinforced polypropylene composite. The composite samples were fabricated with and without the use of PP-grafted maleic anhydride compatibilizer. The fatigue life of the composite sample was improved in the presence of a compatibilizer. Moreover, the fatigue failure mechanism got changed with coir fibre in the polypropylene matrix. The fracture mechanism in the virgin polymer was thermal fatigue, whereas in the composite sample it was mechanical fatigue [78]. In another research work by the same group of authors, coir fibre-reinforced polypropylene composite was prepared by following the design of an experiment in which coir fibre and compatibilizer content were taken as the independent variable. The coir fibre content was found to have the highest effect on mechanical properties like tensile and impact strength. The effect of increasing compatibilizer content was only observed for the elastic modulus [79]. Owolabi and Czvikovszky used chopped CF (coir) to compound polypropylene to prepare irradiated composite material. The coir fibres were irradiated by using a few crosslinking additives to enhance the bonding between matrix polymer and reinforcing fibre. The coir fibre content was varied from 10% to 50% and a good amount of tensile and impact strength were achieved for the fibre content of 30% [80].

Siddika et al. fabricated a hybrid polypropylene composite and investigated its mechanical properties. The reinforcing material consists of jute and coir fibres with chemical treatment done on the fibre surface. The composite sample displayed the falling tendency of tensile strength with higher filler content. However, the reverse trend in elastic modulus variation was observed with filler loading. Other mechanical properties like impact strength, hardness value, and flexural strengths were perceived to be enhanced with fibre content [81]. Zaman and Beg prepared the coir fibre-reinforced polypropylene composite using the compression moulding technique. The effect of fibre content with surface treatment of fibre using stearic acid on mechanical properties of fabricated composite sample was investigated. The impact strength, elastic modulus, and tensile strength of composite material were enhanced with the addition of fibre content. The surface modification with stearic acid has also resulted in improved mechanical strengths [82]. In another research work, polypropylene matrix was reinforced with unidirectional coir fibre to prepare compression-moulded composite material. The fibre surface was modified using alkali treatment followed by tetra-methoxy orthosilicate to enhance adhesion with polymer matrix. The mechanical properties got improved with an increase in fibre content and the presence of chemical treatment. The fibre loading of 30 wt.% was found to display optimum mechanical properties [83].

7.3.2 Rubber-based composites

Albano et al. prepared both treated and untreated coconut flour-reinforced nitrile rubber composite. The thermal degradation behaviour of fabricated composite material was investigated. The addition of coconut flour resulted in a substantial enhancement in the thermal stability of rubber (E_{inv} value increased from 89 kJ/mol to 277 kJ/mol). E_{inv} represents the invariant activation energy that represents the minimum energy required to start a reaction. Furthermore, the E_{inv} value for treated coconut flour composite was enhanced to 298 kJ/mol. Nevertheless, the maximum temperature of decomposition remained constant irrespective of filler addition [84]. Geethamma et al. analysed the reinforcing effect of short coir fibres on the stress relaxation performance of natural rubber. The entropy of the material can be quantified by the rate of stress relaxation. The stress relaxation rate was substantially affected by the strain level at a later one's higher value [85]. In another research work, the same group of authors has prepared untreated and treated coir fibre-reinforced natural rubber composites. The coir fibres of various lengths like 6 mm, 10 mm, and 14 mm were used for reinforcement purposes. Moreover, 10-mm long coir fibre was subjected to surface treatment using 5 wt.% NaOH solution for 4, 24, 48, and 72 hours. The stress–strain behaviour of rubber was found to be improved by the inclusion of fibre with alkali treatment [86]. Herrera et al. investigated the reinforcing impact of coconut flour particles on the nitrile butadiene rubber (NBR) by analysing its vulcanization behaviour. The flour particle amount used for reinforcement was 20 phr. The curing reaction of NBR was affected by the presence of agglomerates of coconut flour particle (size less than 300 μm) as evident from vulcanization kinetics [87]. Sareena et al. focused on the utilization of treated and untreated coconut shell powder as reinforcing filler in natural rubber to prepare composite by an open mill-mixing technique. The reinforcing effect of coconut filler was observed to be a maximum of 10 parts per hundred loadings. Moreover, treated filler showed greater reinforcing ability and better physical–mechanical properties as compared to untreated filler [88].

Wei and Gu fabricated a coir fibre-reinforced rubber-based composite board and investigated the variation of its tensile strength. The heat press technique was used for composite board fabrication. The fibre volume of 60% was found to be showing maximum tensile strength. Any alteration in fibre volume from 60% (be it greater or smaller) resulted in adversely affecting the tensile strength [2]. Arumugam et al. reinforced natural rubber with raw and treated CF by vulcanizing at 153°C. The properties of the prepared composite were investigated by stress–strain analysis along with hardness and abrasion wear test. Moreover, bonding agents were used to improve the adhesion and bonding between reinforcing fibres and rubber [89].

7.3.3 Cement-based composite

Andic-Cakir et al. prepared coir fibre-reinforced fine aggregate-based cement composite. The coir fibre wt.% was 0.4%, 0.6%, and 0.75% of total mixture weight. The composite material was investigated for mechanical properties, water absorption ability, and thermal conductivity. Furthermore, alkali-treated fibres were also used for reinforcement in the same ratio to prepare cement composite. The same set of properties was also assessed for treated fibre composite. The fibre inclusion in the cement matrix along with fine aggregate improved its thermal and mechanical properties. However, their unit weight got reduced. With the increase in fibre amount and chemical treatment, these properties variation turn out to be more prominent [90]. Cook et al. utilized short coir fibres as reinforcement in the cement matrix. They aimed for the application of developed composite in the area of roof materials. The fibre distribution in the matrix was random. The vol.% and length of fibres along with casting pressure were selected as basic parameters for composite fabrication. The dimensional stability, water uptake, combustibility, and mechanical properties were investigated for the developed material. The experimental results showed the 37.5-mm fibre length and 7.5% fibre volume as the optimum combination of the material parameter. The properties along with the cost of material were compared with existing roof material like corrugated galvanized iron or asbestos–cement sheet and found the previous one better [91].

7.4 APPLICATION AND FUTURE PROSPECTS

The application of coir fibres and its polymer composites consists of traditional as well as new emerging areas. The traditional area of coir fibres applications are rope, doormats, brushes, mattresses, blankets, and so forth. The emergent zone of applications of coir fibre-reinforced plastics depicts the thrilling significance in every field, from polymer industries to agriculture to the automobile sector to real estate, thus utilizing its hidden potential in all possible ways [92]. The prospective applications of coir fibre-reinforced polymers are summarized and illustrated in Table 7.4. Furthermore, the economic feasibility and technical viability of coir-based composites can lead to explore

Table 7.4 Application of coir fibre-reinforced polymer composites

Coir fibre-reinforced composites	Application	Reference
Hybrid particle board composites made from rice straws and coir fibres	Particle boards	[93]
Coir fibre–cementitious composites	Sound absorption panel	[94]
Coir fibre-reinforced concrete beams externally strengthened with flax FRP composites	Packaging material	[95]
Coir fibre-reinforced polylactic acid	Lightweight structural components like panels, decks, and slabs	[22]
Polylactic acid (PLA) biocomposites reinforced with coir fibres	Seat cushions in automobiles	[96]

novel applications like consumer goods, low-cost housing, boat hulls, and storage tanks, in the aerospace industry (propellers, wings, and tails). Other relevant applications should be in form of household products (as replacement of wood-based goods), decorating items for indoor, sole material for shoes and sandals, and lightweight sports items. Moreover, the fabrication of partition boards and false ceilings using these materials are other budding application areas that have much potential. Various applications suggested by several researchers are particle boards [93], cement mortar, sound absorption panel [94], packaging material [95], lightweight structural components like panels, decks, and slabs [22], and seat cushions in automobiles [96].

This chapter considered the integrated effects of coir fibres on the partial and fully biodegradable composites on its mechanical, thermal, and physical properties. The future prospects of coir fibre utilization can be in the area of improving interfacial adhesion/bonding between polymer and coir fibre. Another sincere effort is required to address the issues of voids/pores in coir fibre composites. Moreover, one of the major hindrances in the application of coir fibre composite is the absorption of moisture; therefore, this can become a foremost domain of investigation for researchers. Coconut coir fibres are the versatile kind of tough and weather-resistant natural fibre. Through optimal use of surface modification, it can be utilized as reinforcement in polymers/non-polymers to be used as industrial material subjected to various end-use applications. To boost the physical, mechanical, and thermal properties of the coir filler or fibre can be extended to next level by developing hybrid reinforcement composite materials. The matrix plays a major role in binding the reinforcement to develop strength; however, the brittle response of thermosetting polymer quickly propagates the crack growth. In order to hinder the crack initiation and crack growth, a suitable proportion of polymer matrix can be suggested for developing composite structures.

7.5 CONCLUSIONS

An overview of the developments made in the area of coir fibre-reinforced composites is presented in terms of their mechanical properties, manufacturing procedures, and applications. Major conclusions drawn from the literature review are as follows:

- The mechanical, thermal, physical, and tribological properties of the coir fibre-reinforced polymer composite invariably depend on coir fibre content in the composite.
- For most of the coir fibre composite, mechanical properties got enhanced up to a certain amount of coir fibre content, and beyond that there is a consistent drop in properties.
- The coconut reinforcement in thermoplastic and thermoset polymer is utilized in three forms: long fibre, short fibre, and shell flour filler.

- Various surface modification techniques are employed by several researchers like alkali treatment, silane treatment, *o*-hydroxybenzene diazonium salt, and bromododecane.
- Exploring innovative applications, such as consumer goods, low-cost housing, boat hulls, storage tanks, within the aerospace industry (propellers, wings, and tails) can be facilitated by the economic feasibility and technical viability of coir-based composites.

ACKNOWLEDGEMENTS

The authors would like to thank the Department of Mechanical Engineering, School of Engineering, Dayananda Sagar University Bangalore, India, for giving essential research facilities and a conducive environment without which the present research work could never been completed.

FUNDING

No funding was received to assist with the preparation of this chapter.

CONFLICT OF INTEREST

The author(s) declare no potential conflicts of interest with respect to the research, authorship, and/or publication of this chapter.

CONSENT TO PARTICIPATE

Informed consent was obtained from participants included in the study.

CONSENT FOR PUBLICATION

Not applicable.

AVAILABILITY OF DATA AND MATERIALS

Not applicable as figures are taken from an open source that falls under a Creative Commons Attribution License.

AUTHORS CONTRIBUTION

Rahul Kumar: Conceptualization, writing, review and editing. **Faladrum Sharma:** writing, review and editing. **Sumit Bhowmik:** Supervising, reviewing, editing and fine-tuning. **P M G Bashir Asdaque:** review and editing.

REFERENCES

1. UNEP (2011) Towards a green economy: Pathways to sustainable development and poverty eradication—A synthesis for policy makers. www.unep.org/greeneconomy (accessed on September 13, 2021).
2. Wang W, Huang G (2009) Characterisation and utilization of natural coconut fibres composites. Mater Des 30: 2741–2744
3. Salazar VLP, Leão AL, Rosa DS, Gomez JGC, Alli RCP (2011) Biodegradation of coir and sisal applied in the automotive industry. J Polym Environ 19: 677–688.
4. Kumar R, Bhowmik S (2021) Quantitative probing of static and dynamic mechanical properties of different bio-filler-reinforced epoxy composite under assorted constraints. Polym Bull 78: 1231–1252.
5. Kumar R, Bhowmik S, Kumar K, Davim JP (2020) Perspective on the mechanical response of pineapple leaf filler/toughened epoxy composites under diverse constraints. Polym Bull 77: 4105–4129.
6. Iovino R, Zullo R, Rao MA, Cassar L, Gianfreda L (2008) Biodegradation of poly(lactic acid)/starch/coir biocomposites under controlled composting conditions. Polym Degrad Stab 93: 147–157.
7. Jang JY, Jeong TK, Oh HJ, Youn JR, Song YS (2012) Thermal stability and flammability of coconut fiber reinforced poly(lactic acid) composites. Compos B Eng 43: 2434–2438.
8. Dong Y, Ghataura A, Takagi H, Haroosh HJ, Nakagaito AN, Lau KT (2014) Polylactic acid (PLA) biocomposites reinforced with coir fibres: Evaluation of mechanical performance and multifunctional properties. Compos A: Appl Sci Manuf 63: 76–84.
9. Kumar R, Kumar K, Bhowmik S, Sarkhel G (2019) Tailoring the performance of bamboo filler reinforced epoxy composite: Insights into fracture properties and fracture mechanism. J Polym Res 26: 54–68.
10. Akay, O., Altinkok, C., Acik, G., Yuce, H., Ege, G. K. (2021). A bio-based and non-toxic polyurethane film derived from *Luffa cylindrica* cellulose and ʟ-lysine diisocyanate ethyl ester. Eur Polym J 161, 110856.
11. Kumar R, Kumar K, Bhowmik S (2018) Mechanical characterization and quantification of tensile, fracture and viscoelastic characteristics of wood filler reinforced epoxy composite. Wood Sci Technol 52: 677–699.
12. Diao C (2015) Wheat gluten biodegradable plastics and biocomposites.
13. Hemsri S, Grieco K, Asandei AD, Parnas RS (2012) Wheat gluten composites reinforced with coconut fiber. Compos A: Appl Sci Manuf 43: 1160–1168.
14. Corradini E, Imam SH, Agnelli JA, Mattoso LH (2009) Effect of coconut, sisal and jute fibers on the properties of starch/gluten/glycerol matrix. J Polym Environ 17: 1–9.
15. Ramírez MGL, Satyanarayana KG, Iwakiri S, de Muniz GB, Tanobe V, Flores-Sahagun TS (2011) Study of the properties of biocomposites. Part I. Cassava starch-green coir fibers from Brazil. Carbohydr Polym 86: 1712–1722.
16. Nam TH, Ogihara S, Tung NH, Kobayashi S (2011) Effect of alkali treatment on interfacial and mechanical properties of coir fiber reinforced poly(butylene succinate) biodegradable composites. Compos B Eng 42: 1648–1656.
17. Adeniyi AG, Onifade DV, Ighalo JO, Adeoye AS (2019) A review of coir fiber reinforced polymer composites. Compos B Eng 176, 107305.
18. Hasan, KF, Horváth, PG, Bak, M, Alpár, T (2021) A state-of-the-art review on coir fiber-reinforced biocomposites. RSC Adv 11(18), 10548–10571.
19. Jayavani, S, Deka, H, Varghese, TO, Nayak, SK (2016). Recent development and future trends in coir fiber reinforced green polymer composites: Review and evaluation. Polym Compos 11(37), 3296–3309.

20. Asasutjarit C, Charoenvai S, Hirunlabh J, Khedari J (2009) Materials and mechanical properties of pretreated coir-based green composites. Compos B Eng 40: 633–637.

21. Rout J, Misra M, Tripathy SS, Nayak SK, Mohanty AK (2001) Novel eco-friendly biodegradable coir-polyester amide biocomposites: Fabrication and properties evaluation. Polym Compos 22: 770–778.

22. Siakeng R, Jawaid M, Ariffin H, Salit MS (2018) Effects of surface treatments on tensile, thermal and fibre-matrix bond strength of coir and pineapple leaf fibres with poly lactic acid. J Bionics Eng 15:1035–1046.

23. Perez-Fonseca AA, Arellano M, Rodrigue D, Gonzalez-Núñez R, Robledo-Ortíz JR (2016) Effect of coupling agent content and water absorption on the mechanical properties of coir-agave fibers reinforced polyethylene hybrid composites. Polym Compos 37:3015–24.

24. Abraham E, Deepa B, Pothen LA, Cintil J, Thomas S, John MJ, …, Narine SS (2013) Environmental friendly method for the extraction of coir fibre and isolation of nanofibre. Carbohydr Polym 92: 1477–1483.

25. Kumar R, Kumar K, Bhowmik S (2018) Assessment and response of treated *Cocos nucifera* reinforced toughened epoxy composite towards fracture and viscoelastic properties. J Polym Environ 26: 2522–2535.

26. Priya NAS, Raju PV, Naveen P (2014) Experimental testing of polymer reinforced with coconut coir fiber composites. Int J Emerg Technol Adv Eng 4:453–460.

27. Saw SK, Sarkhel G, Choudhury A (2011) Surface modification of coir fibre involving oxidation of lignins followed by reaction with furfuryl alcohol: Characterization and stability. Appl Surf Sci 257:3763–3769.

28. Verma D, Gope PC, Shandilya A, Gupta A, Maheshwari MK (2013) Coir fibre reinforcement and application in polymer composites. J Mater Environ Sci 4: 263–276.

29. Chun KS, Husseinsyah S, Osman H (2013) Properties of coconut shell powder-filled polylactic acid ecocomposites: Effect of maleic acid. Polym Eng Sci 53: 1109–1116.

30. Harish S, Michael DP, Bensely A, Lal, DM, Rajadurai A (2009) Mechanical property evaluation of natural fiber coir composite. Mater. Charact 60: 44–49.

31. Biswas S, Kindo S, Patnaik A (2011) Effect of fiber length on mechanical behavior of coir fiber reinforced epoxy composites. Fiber Polym 12: 73–78.

32. Sarki J, Hassan SB, Aigbodion VS, Oghenevweta JE (2011). Potential of using coconut shell particle fillers in eco-composite materials. J Alloys Compd 509: 2381–2385.

33. Aireddy H, Mishra SC (2011) Tribological behavior and mechanical properties of bio-waste reinforced polymer matrix composites. J Metall Mater Sci 53: 139–152.

34. Saw SK, Sarkhel G, Choudhury A (2012) Preparation and characterization of chemically modified jute–coir hybrid fiber reinforced epoxy Novolac composites. J Appl Polym Sci 125: 3038–3049.

35. Bhagat VK, Biswas S, Dehury J (2014) Physical, mechanical, and water absorption behavior of coir/glass fiber reinforced epoxy based hybrid composites. Polym Compos 35: 925–930.

36. Narendar R, Dasan KP, Nair M (2014) Development of coir pith/nylon fabric/epoxy hybrid composites: Mechanical and ageing studies. Mater Des 54: 644–651.

37. Gope PC, Rao, DK (2016) Fracture behaviour of epoxy biocomposite reinforced with short coconut fibres (*Cocos nucifera*) and walnut particles (*Juglans regia* L.). J. Thermoplast Compos Mater 29: 1098–1117.

38. Kumar SMS, Duraibabu D, Subramanian K (2014) Studies on mechanical, thermal and dynamic mechanical properties of untreated (raw) and treated coconut sheath fiber reinforced epoxy composites. Mater Des 59: 63–69.

39. Ojha S, Raghavendra G, Acharya SK (2014) A comparative investigation of bio waste filler (wood apple-coconut) reinforced polymer composites. Polym Compos 35: 80–185.
40. Prasad GLE, Gowda BSK, Velmurugan R (2014) Prediction of properties of coir fiber reinforced composite by ANN. Experimental Mechanics of Composite, Hybrid, and Multifunctional Materials 6: 1–7.
41. Monteiro SN, Terrones LAH, D'Almeida JRM (2008) Mechanical performance of coir fiber/polyester composites. Polym Test 27: 591–595.
42. Rout J, Misra M, Mohanty AK, Nayak SK, Tripathy SS (2003) SEM observations of the fractured surfaces of coir composites. J Reinf Plast Compos 22: 1083–1100.
43. Rout J, Misra M, Tripathy SS, Nayak SK, Mohanty AK (2001) The influence of fibre treatment on the performance of coir–polyester composites. Comp Sci Technol 61: 1303–1310.
44. Prasad SV, Pavithram C, Rohatgi PK (1983) Alkali treatment of coir fibres for coir–polyester composites. J Mater Sci 18: 1443–1454.
45. Hill CAS, Khalil HPSA (2000) The effect of environmental exposure upon the mechanical properties of coir or oil palm fiber reinforced composites. J Appl Polym Sci 77: 1322–1330.
46. Hill CAS, Khalil HPSA (2000) Effect of fiber treatments on mechanical properties of coir or oil palm fiber reinforced polyester composites. J Appl Polym Sci 78: 1685–1697.
47. Mulinari DR, Baptista CARP, Souza JVC, Voorwald HJC (2011) Mechanical properties of coconut fibers reinforced polyester composites. Procedia Eng 10: 2074–2079.
48. Jayabal S, Rajamuneeswaran S, Ramprasath R, Balaji NS (2013) Artificial neural network modeling of mechanical properties of calcium carbonate impregnated coir-polyester composites. Trans Indian Inst Met 66:247–255.
49. Jayabal S, Velumani S, Navaneethakrishnan P, Palanikumar K (2013) Mechanical and machinability behaviors of woven coir fiber-reinforced polyester composite. Fibers Polym 14: 1505–1514.
50. Jayabal S, Natarajan U (2011) Influence of fiber parameters on tensile, flexural, and impact properties of nonwoven coir–polyester composites. Int J Adv Manuf Syst 54: 639–648.
51. Yousif BF (2009) Frictional and wear performance of polyester composites based on coir fibres. Proc Inst Mech Eng J Eng Tribo 223: 51–59.
52. Owolabi O, Czvikovszky T, Kovács I (1985) Coconut-fiber-reinforced thermosetting plastics. J Appl Polym Sci 30: 1827–1836.
53. Rout J, Tripathy SS, Misra M, Mohanty AK, Nayak SK (2001) The influence of fiber surface modification on the mechanical properties of coir–polyester composites, Polym Compos 22: 468–476.
54. Abdullah NM, Ahmad I (2013) Potential of using polyester reinforced coconut fiber composites derived from recycling polyethylene terephthalate (PET) waste. Fiber Polym 14: 584–590.
55. Jappes JTW, Siva I, Rajini N (2012) Fractography analysis of naturally woven coconut sheath reinforced polyester composite: A novel reinforcement. Polym Plast Technol Eng 51: 419–424.
56. Brahma KM, Pavithran C, Pilai, RM (2005) Coconut fibre reinforced polyethylene composites: Effect of natural waxy surface layer of the fibre on fibre/matrix interfacial bonding and strength of composites. Compos Sci Technol 65:563–569.
57. Arrakhiz FZ, AchabyME, Kakou AC, Vaudreuil S, Benmoussa K, Bouhfid R, Fassi-Fehri O, Qaiss A (2012) Mechanical properties of high density polyethylene reinforced with chemically modified coir fibers: Impact of chemical treatments. Mater Des 37: 379–383.

58. Albano C, González J, Hernández M, Ichazo MN, Alvarado S, Ziegler DM, Lavadi M (2009) HDPE-coconut flour composites: Effect of coupling agents and surface modification, Macromolecular Symposia, 286: 70–80.

59. Fernandes EM, Correlo VM, Mano JF, Reis RL (2013) Novel cork–polymer composites reinforced with short natural coconut fibres: Effect of fibre loading and coupling agent addition. Compos Sci Tech 78: 56–62.

60. Leblanc JL (2006) Poly(vinyl chloride)–green coconut fiber composites and their nonlinear viscoelastic behavior as examined with Fourier transform rheometry. J Appl Poly Sci 101: 3638–3651.

61. Leblanc JL, Furtado CRG, Leite MCAM, Visconte LLY, de Souza AMF (2007) Effect of the fiber content and plasticizer type on the rheological and mechanical properties of poly(vinyl chloride)/green coconut fiber composites. J Appl Poly Sci 106: 3653–3665.

62. Tran LQN, Fuentes CA, Dupont-Gillain C, Vuure AWV, Verpoest I (2013) Understanding the interfacial compatibility and adhesion of natural coir fibre thermoplastic composites, Compos Sci Tech 80: 23–30.

63. Leblanc JL, Furtado CRG, Leite MCAM, Visconte LLY, Ishizaki MH (2006) Investigating polypropylene–green coconut fiber composites in the molten and solid states through various techniques. J Appl Poly Sci 102:1922–1936.

64. Rozman HD, Tay GS, Kumar RN, Abubakar A, Ismail H, Ishak ZAM (1999) Polypropylene hybrid composites: A preliminary study on the use of glass and coconut fiber as reinforcements in polypropylene composites. PolymPlast Technol Eng 38: 997–1011.

65. Sharma SC, Krishna M, Narasimhamurthy HN, Murthy S (2006) Studies on the weathering behavior of glass coir polypropylene composites. J Reinf Plast. Compos. 25: 9925–932.

66. Lai CY, Sapuan SM, Ahmad M, Yahya N (2005) Mechanical and electrical properties of coconut coir fibre-reinforced polypropylene composites. PolymPlast Technol Eng 44: 619–632.

67. Haque MM, Hasan M, Islam MS, Ali ME (2009) Physico-mechanical properties of chemically treated palm and coir fiber reinforced polypropylene composites. Bioresour Technol 100: 4903–4906.

68. Gu H (2009) Tensile behaviours of the coir fibre and related composites after NaOH treatment. Mater Des 30: 3931–3934.

69. Islam MN, Rahman MR, Haque MM, Huque MM (2010) Physico-mechanical properties of chemically treated coir reinforced polypropylene composites. Compos A 41: 192–198.

70. Ayrilmis N, Jarusombuti S, Fueangvivat V, Bauchongko lP, White RH (2011) Coir fiber reinforced polypropylene composite panel for automotive interior applications. Fiber Polym 12: 919–926.

71. Bettini SHP, Biteli AC, Bonse BC, Morandim-Giannetti AA (2014) Polypropylene composites reinforced with untreated and chemically treated coir: Effect of the presence of compatibilizer. Polym Eng. Sci. 55:2050–2057. DOI: 10.1002/pen.24047.

72. Morandim-Giannetti AA, Agnelli JAM., Lancas BZ, Magnabosco R, Casarin SA, Bettini SHP (2012) Lignin as additive in polypropylene/coir composites: Thermal, mechanical and morphological properties. Carbohydr. Polym. 87: 2563–2568.

73. Haque MM, Islam MS, Islam MN (2012) Preparation and characterization of polypropylene composites reinforced with chemically treated coir. J Polym Res 19:9847.

74. Arrakhiz FZ, Malha M, Bouhfid R, Benmoussa K, Qaiss A (2013) Tensile, flexural and torsional properties of chemically treated alfa, coir and bagasse reinforced polypropylene. Compos: Part B 47: 35–41.

75. Nandi A, Kale A, Raghu N, Aggarwal PK, Chauhan SS (2013) Effect of concentration of coupling agent on mechanical properties of coir–polypropylene composite. J Indian Acad Wood Sci 10:62–67.
76. Mir SS, Nafsin N, Hasan M, Hasan N, Hassan A (2013) Improvement of physico-mechanical properties of coir-polypropylene biocomposites by fiberchemical treatment. Mater Des 52: 251–257.
77. Zaman HU, Khan MA, Khan RA (2012) Comparative experimental measurements of jute fiber/polypropylene and coir fiber/polypropylene composites as ionizing radiation. Poly Compos 33: 1077–1084.
78. Bettini SHP, Antunes MC, Magnabosco R (2011) Investigation on the effect of a compatibilizer on the fatigue behavior of PP/coir fiber composites. Polym Eng Sci 51: 2184–2190.
79. Bettini HP, Bicudo ABLC, Augusto IS, Antunes LA, Morassi PL, Condotta R, Bonse BC (2010) Investigation on the use of coir fiber as alternative reinforcement in polypropylene. J Appl Polym Sci 118: 2841–2848.
80. Owolabi O, Czvikovszky T (1988) Composite materials of radiation-treated coconut fiber and thermoplastics. J Appl Polym Sci 35: 573–582.
81. Siddika S, Mansura F, Hasan M, Hassan A (2014) Effect of reinforcement and chemical treatment of fiber on the properties of jute-coir fiber reinforced hybrid polypropylene composites. Fiber Polym 15: 1023–1028.
82. Zaman HU, Beg MDH (2014) Effect of coir fiber content and compatibilizer on the properties of unidirectional coir fiber/polypropylene composites. Fiber Polym 15: 831–838.
83. Zaman HU, Beg MDH (2014) Preparation, structure, and properties of the coir fiber/polypropylene composites. J Compos Mater 48: 3293–3301.
84. Albano C, Ichazo MN, Boyer I, Hernández M, González J, Karama A, Covis M (2012) Study of the thermal stability of nitrile rubber-coconut flour compounds. Polym Degrad Stab 97: 2202–2211.
85. Geethamma VG, Pothen LA, Rhao B, Neelakantan NR, Thomas S (2004) Tensile stress relaxation of short-coir-fiber-reinforced natural rubber composites. J Appl Polym Sci 94: 96–104.
86. Geethamma VG, Joseph R, Thomas S (1995) Short coir fiber-reinforced natural rubber composites: Effects of fiber length, orientation, and alkali treatment. J Appl Polym Sci 55: 583–594.
87. Herrera R, Ichazo MN, Albano C, Hernández M, González J (2011) Curing kinetics of NBR filled with coconut flour. Polym Compos 32: 529–536.
88. Sareena C, Ramesan MT, Purushothaman E (2012) Utilization of coconut shell powder as a novel filler in natural rubber. J Reinf Plast Compos 31: 8533–8547.
89. Arumugam N, Tamare Selvy K, Rao KV, Rajalingam P (1989) Coconut-fiber-reinforced rubber composites. J Appl Polym Sci 37: 2645–2659.
90. Andic-Cakir O, Sarikanat M, Tufekci HB, Demirci C, Erdoğan UH (2014) Physical and mechanical properties of randomly oriented coir fiber–cementitious composites. Compos B Eng 61: 49–54.
91. Cook DJ, Pama RP, Weerasingle HLSD (1978) Coir fibre reinforced cement as a low cost roofing material. Build Environ 13:193–198.
92. Kumar R, Bhowmik S (2019) Elucidating the coir particle filler interaction in epoxy polymer composites at low strain rate. Fibers Polym 20:428–439.
93. Zhang L, Hu Y (2014) Novel lignocellulosic hybrid particleboard composites made from rice straws and coir fibers. Mater Des 55:19–26.
94. Andiç-Çakir ÇO, Sarikanat M, Tüfekçi HB, Demirci C, Erdogan ÜH (2014) Physical and mechanical properties of randomly oriented coir fiber–cementitious composites. Compos B Eng 61:49–54.

95. Yan L, Su S, Chouw N (2015) Microstructure, flexural properties and durability of coir fibre reinforced concrete beams externally strengthened with flax FRP composites. Compos B Eng 80:343–354.
96. Dong Y, Ghataura A, Takagi H, Haroosh HJ, Nakagaito AN, Lau KT (2014) Polylactic acid (PLA) biocomposites reinforced with coir fibres: Evaluation of mechanical performance and multifunctional properties. Compos Appl Sci Manuf 63: 76–84.
97. Yan L, Chouw N, Huang L, Kasal B (2016) Effect of alkali treatment on microstructure and mechanical properties of coir fibres, coir fibre reinforced-polymer composites and reinforced-cementitious composites. Constr Build Mater 112:168–182.
98. Yusoff RB, Takagi H, Nakagaito AN (2016) Tensile and flexural properties of polylactic acid-based hybrid green composites reinforced by kenaf, bamboo and coir fibers. Ind Crops Prod 94:562–573.
99. Sanjay M, Madhu P, Jawaid M, Senthamaraikannan P, Senthil S, Pradeep S (2018) Characterization and properties of natural fiber polymer composites: A comprehensive review. J Clean Prod 172:566–581.
100. Balaji NS, Chockalingam S, Ashokraj S, Simson D, Jayabal S (2020) Study of mechanical and thermal behaviours of zea-coir hybrid polyester composites. Mater Today Proc. 27:2048–2051.
101. dos Santos JC, de Oliveira LÁ, Vieira LMG, Mano V, Freire RT, Panzera TH (2019) Eco-friendly sodium bicarbonate treatment and its effect on epoxy and polyester coir fibre composites. Constr Build Mater 211:427–436.
102. Thomas S, Paul SA, Pothan LA, Deepa B (2011) Natural fibres: Structure, properties and applications. In Cellulose Fibers: Bio-and Nano-polymer Composites. Springer, Berlin, Heidelberg, pp. 3–43.
103. Centre Coir Research Board, GOI (2021) http://coirboard.gov.in/?page_id=74 (accessed 27 September 2021).
104. Sharma F, Kumar R, Bhowmik S (2023). Response of coconut coir filler-reinforced epoxy composite toward cyclic loading: Fatigue property evaluation. In: Joshi SN, Dixit US, Mittal RK, Bag S (eds) Low Cost Manufacturing Technologies. NERC 2022. Springer, Singapore.
105. Aminudin E, Khalid NHA, Azman NA, Bakri K, Din MFM, Zakaria R, Zainuddin N A (2017) Utilization of baggase waste based materials as improvement for thermal insulation of cement brick. In MATEC Web of Conferences (Vol. 103, p. 01019). EDP Sciences.
106. Zimniewska M, Wladyka-Przybylak M. (2016) Natural fibers for composite applications. In: Rana S, Fangueiro R (eds) Fibrous and Textile Materials for Composite Applications: Textile Science and Clothing Technology. Springer, Singapore.

Chapter 8

Epoxy-based corrosion-resistant coating for marine engineering application: Processing principles, and applications

Rajan Kumar, Nachiketa Das, Sumit K. Sharma, and Amarish Kumar Shukla

8.1 INTRODUCTION

For decades, corrosion has presented a significant challenge and a considerable threat to society and the economy. Around one-third of the world's metal production is lost due to corrosion in technical applications. This corrosion-related loss accounts for an estimated 7–8% of the gross national product (GNP) of industrialized nations. The issue has intensified with rapid development, industrialization, and the construction of large infrastructures, all exposing vulnerable metals to polluted environments. Steel, the most widely used metal globally, is particularly prone to degradation from environmental corrosion [1]. This encompasses the annual direct expenses attributed to metallic corrosion in various sectors such as infrastructure (including material storage, pipelines, and highway bridges), government defense, automotive transportation, and utilities (including gas/electric and drinking water) [2]. Marine platforms like ships, offshore structures, and steel bridges face increased susceptibility to corrosion when exposed to seawater environments. Ships larger than 100 m in length are predominantly constructed using mild steel (MS) for their primary structure, with fast ferries of larger sizes being an exception as they are often built with aluminum. MS is susceptible to rapid corrosion in marine environments, necessitating extensive measures to protect ships from this corrosive effect [3]. Significant losses resulting from material degradation have spurred substantial efforts to develop new corrosion prevention strategies. Several mitigation measures have been implemented to control corrosion, including organic/polymer coatings, corrosion inhibitors, and anodic/cathodic protection, which find widespread use in marine, automobile, construction, and pipeline industries. Among these methods, the latest classifications of corrosion-protective coatings are gaining popularity due to their ease and scalability of application. Various metal structures are being protected, and their stability and lifespan are extended by applying organic polymer coatings, composite coatings, hydrophobic coatings, and self-cleaning coatings [4].

Epoxy technology is extensively employed as a coating material across various sectors, such as automotive, piping, and electronic equipment encapsulation. It

126

DOI: 10.1201/9781003564355-8

serves as a substitute for casting in aircraft and automated machinery applications. Additionally, composite materials are widely used in industries like aerospace, printed circuit board (PCB) manufacturing, and the production of pressure pipes [5]. The term "epoxy" is a prefix that denotes a compound comprising an oxygen atom bonded to two other atoms. An epoxy resin encompasses any molecule containing one or more 1,2-epoxy groups. Cyclohexene oxide is alternatively known as 1,2-epoxycyclohexene, 1-2-oxidocyclohexene, and 7-oxabicycloheptane. When epoxides are viewed as derivatives of α-glycols, the prefix "hydro" is employed with the convenient term "glycidyl". Several simple epoxides have common names corresponding to their structures; for instance, 1-chloro-2,3-epoxypropane, 1,2-epoxy-3-hydroxypropane, and 2,3-epoxypropanoic acid are referred to as epichlorohydrin, glycidol, and glycidic acid, respectively [6]. The properties of polymer nanocomposites can be influenced independently or simultaneously by factors such as the shape, size, and quantity of nanoparticles [7, 8]. In the past decade, epoxy resins incorporating nanofillers have emerged in nanotechnology. These advanced materials offer exceptional properties, making them ideal for high-performance adhesives in demanding environments such as exterior car body panels, aircraft, and marine applications [9–11]. Epoxy systems modified with nanoclay are commonly utilized in various applications, including tooling, laminates, molding, casting, and particularly in the construction sector [12]. From a technological standpoint, many epoxy nanocomposites have entered the market as paints and coatings. Numerous studies have highlighted the advantageous characteristics of these materials, including notable improvements in mechanical and chemical resistance and providing a pleasing decorative finish [13]. Several literatures have emphasized the superior properties of nanostructured epoxy coating systems, such as self-healing capabilities, microwave absorbance, and enhanced heat resistance [14–18]. Epoxy, a traditional thermoset resin (a plastic material initially a liquid monomer or a prepolymer), has become a staple in numerous industrial coating formulations. Through the integration of nanotechnology, epoxy has exhibited a wide range of advantageous properties. Incorporating nanoparticles into the epoxy matrix has enabled the development of diverse epoxy nanocomposite coatings tailored for specific applications across various conditions [19]. Epoxy is renowned for its ability to coat surfaces effectively, thereby preventing corrosion on metallic surfaces. The epoxy shows high corrosion resistance because of its exceptional chemical stability, excellent corrosion resistance, high tensile strength, minimal shrinkage during curing, and superior adhesion properties [20]. Engineers seek smart materials for applications where long-term competitiveness relies on superior properties. Epoxy is the most favored thermoset material, characterized by resins formed through step-wise or chain-wise polymerization reactions, and is extensively utilized in the coating industry. To ensure high efficiency, it is crucial to thoroughly comprehend the interplay among curing, properties, and performance of epoxy coatings. Epoxy nanocomposite coatings represent a typical application of thermoset resins fortified with nanoparticles for engineering purposes. Irrespective of the chemistry of epoxy resins, their functionality, and the lubricity of the system, the coating industry has significantly benefited from the promising features enabled by incorporating nanoparticles into epoxy formulations [21]. Epoxy resin has versatility and practical attributes encompassing physical, chemical, and mechanical properties, safety, and cost-effectiveness. Because of that, it is one of the premier choices among various materials utilized for paint coatings. Despite the availability of numerous coating systems based on epoxy resins, they have yet to establish themselves in field applications, necessitating high fouling resistance. Nanostructured epoxy materials have gained recognition in aerospace and marine coatings for their anticorrosive and self-healing capabilities, effectively limiting the intrusion of aggressive ions. Moreover, advanced antifouling and antibacterial epoxy nanocomposite coatings, characterized by their environment-friendly properties, have been utilized on the hulls of ships and marine structures [22].

8.2 CORROSION IN SHIP STRUCTURES AND ITS PREVENTION

The term corrosion describes the deterioration of materials over time as a result of chemical reactions with the environment, which causes components to stop working properly. It entails permanent surface damage to metal, which, when subjected to a corrosive environment—which can be solid, liquid, or gas in nature—converts pure metal into chemically more stable forms, such as sulfides, oxides, or hydroxides [23]. Metals must be transformed into their free state using a substantial amount of energy during the extraction process from their ores. They can resist their inclination to exist in merged forms because of this energy. But when these metals come into contact with outside elements like oxygen or moisture, they usually corrode back to their mixed condition. For example, iron corrodes due to environmental factors and forms hydrated ferrous oxide, which is brown. This demonstrates how the process of corrosion works against removing metals from their ores [23]. Figure 8.1 shows the corrosion of ships in a marine environment.

Marine corrosion prevention includes material selection, design considerations, proper usage, and maintenance practices. Metals and nonmetals undergo degradation not solely because of ocean water but also due to various contributing factors associated with seawater. Despite the absence of oxygen, these deposits often create highly corrosive environments. Numerous techniques exist for preventing marine corrosion, including painting, coating application, and corrosion inhibitors. Additionally, coatings and composite structures are highly vulnerable to rapid corrosion. Furthermore, sulfate bacteria found in silt or mud can increase hydrogen sulfide concentration, posing an attack on steel and copper-based alloys. There are various factors, such as natural seawater, stored or recirculated water, brackish coastal seawater,

Figure 8.1 The corroded ship structures [24].

polluted seawater, and synthetic solutions, influencing marine corrosion [25], as shown in Figure 8.2.

Additionally, there are several ways to control corrosion: (a) coating metal surfaces with protective materials, which act as a barrier or sacrificial protection; (b) adding chemicals to the environment to prevent corrosion; (c) changing the composition of alloys to increase resistance; (d) treating metal

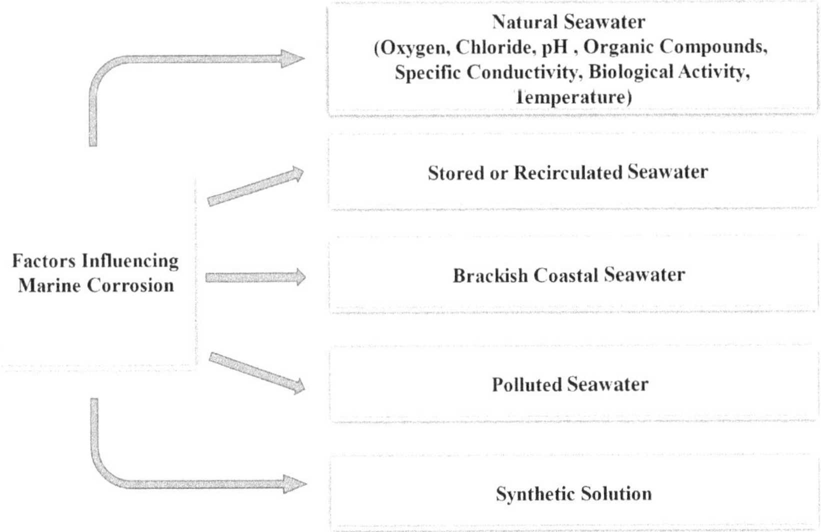

Figure 8.2 Some factors influencing marine corrosion.

surfaces to increase corrosion resistance. Organic coatings serve as the primary defense against corrosion, with complex coating systems utilized in automotive and aerospace industries comprising multiple layers for enhanced protection. These coatings effectively isolate the metal from the environment, emphasizing its barrier properties. Certain additives, like chromates, phosphates, nitrates, molybdates, and organic compounds, are strategically introduced into electrolytes to slow corrosion by preferentially migrating to anodic and/or cathodic sites. Such corrosion inhibitors are widely utilized in piping systems within power plants, chemical processing facilities, and the oil and gas industries [26].

8.3 PRINCIPLES OF EPOXY COATINGS FOR CORROSION PROTECTION

Epoxy coatings usually become touch-dry within 3 hours and reach complete curing after seven days at 25°C. The ambient and surface temperatures significantly affect the duration of complete curing during the curing process. The curing reaction significantly decelerates at temperatures below 10°C. Epoxy coating films exhibit robust resistance against various chemicals, making them outstanding choices for anticorrosion coatings. They are prominently employed in controlling corrosion in marine environments [27]. Lin et al. [28] examined the impact of compromised insulating enamel on the corrosion of high chromium (H), duplex (D), and corrosion-resistant (R) duplex stainless steel. They investigated the corrosion behavior of HDR duplex stainless steel in a 3.5% NaCl solution using local electrochemical impedance spectroscopy (LEIS) and micro-morphology analysis. The LEIS impedance remained steady at approximately 7.0×10^3 Ω within ten days for uncoated duplex stainless steel. However, exposed duplex stainless steel shows the damaged area of steel, and the minimum LEIS impedance fluctuated around 6.5×10^3 Ω for 15 days. The area damaged by the coating, initially 1 mm × 10 mm in size, transformed into a circular hole with a diameter of 1 mm. As a result, the exposed duplex stainless steel LEIS impedance increased in the area of damaged coating. The impedance increased quickly as the measurement went from the damaged to the undamaged coating area. The impedance rise was exactly proportionate to the distance from the damaged area. The time needed for the impedance to stabilize increased with coating thickness. After stabilizing, the thicker coating showed a more notable improvement in boosting duplex stainless steel's resistance to corrosion. Nooredeen et al. [29] studied the corrosion behavior of steel panels coated with eco-friendly (ZrO, Mn, VO, and Bi) phosphomolybdate POM/CoFe$_2$O$_4$@ SiO$_2$ nanocomposite coatings. The coated steel sample was immersed in a 3.5% NaCl solution for three weeks, and the results show that incorporating CoFe$_2$O$_4$@SiO$_2$ nanoparticles improved the electrochemical corrosion resistance of the paint composites. The corrosion results of ZrPOM/CoFe$_2$O$_4$@

SiO_2 composite-based paint on stainless steel plates show decreased charge transfer resistance from 53 MΩ cm^2 to 210 kΩ cm^2, remaining substantially higher than the 5.1 kΩ cm^2 of untreated core–shell painted stainless steel plates. Meanwhile, the R_{ct} values of ZrPOM decreased notably from 410 kΩ cm^2 to 140 kΩ cm^2 after the corrosion test, which is considerably lower than the 210 kΩ cm^2 of the Bi- and VO POM/CoFe$_2$O$_4$@SiO$_2$ painted stainless steel plates. The reduction in R_{ct} impedance values observed after the corrosion test across all examined coatings is likely attributed to the infiltration of water molecules and the migration of ions such as chloride through the coatings. This effect is particularly pronounced in the case of POM coatings compared to core–shell composite coatings. This indicates that a capacitive response at the interface of the stainless steel/NaCl solution remains intact. Edvardsen et al. [30] investigated the influence of various electrical current modes, including DC, AC, and single-polarity pulsed DC, on the surface scaling of electrically conductive epoxy/CNF coatings. Electrical resistance measurements were carried out to gather insights into the bulk and surface resistivity of the epoxy/CNF coating. A cylindrical specimen of the epoxy/CNF coating, with a thickness of 11.3 mm and a diameter of 24.6 mm, was employed to assess volume resistivity. Meanwhile, surface resistivity measurements were conducted using a steel-coated sample. Anodic polarization of the epoxy/carbon nanofiber (CNF) coating at +2 and +5 V during AC and DC conditions restricted scale deposition. However, the mechanisms underlying scale inhibition differed between AC and DC polarization. DC polarization at +2 V yielded the least scale buildup without coating degradation and significantly lowered precipitate accumulation compared to the other tested current types and the reference surface without electrical polarization. However, DC polarization of the epoxy/CNF coating at potentials as high as +5 V resulted in coating degradation. However, employing positive DC with a high pulse frequency of 50 Hz effectively prevented degradation. These findings are significant for numerous industries grappling with surface scaling, particularly when chemical scale inhibition is impractical due to health and safety concerns. Alam et al. [31] studied the effect of the deposition of zirconia particles into the epoxy coating. They deposited varying percentages of zirconium nanoparticles (EZr-1, EZr-2, and EZr-3) on the metal surface coated with epoxy coatings. The deposition of zirconium nanoparticles on the metal surface was studied by energy-dispersive X-ray fluorescence spectroscopy. The results show that as the percentage of zirconium nanoparticles in the coating increases from 1 wt.% to 3 wt.%, the appearance and morphology of the coatings undergo alterations. Specifically, the addition of 1 wt.% and 2 wt.% Zr nanoparticles yield a smooth surface devoid of voids. However, with 3 wt.% Zr, pinholes become apparent on the coating surface. The Fourier transform infrared spectroscopy (FTIR) spectra obtained for the coatings reveal the chemical interaction between ZrO_2 nanoparticles and oxygen and nitrogen atoms in the epoxy chains. The concluded result shows that the 2 wt.% (EZr-2) ZrO_2 nanoparticle coatings demonstrated exceptional corrosion resistance, displaying the highest impedance values compared to the other samples, EZr-1 and EZr-3. Afshar et al. [32] studied MS reinforcement with hot-dip galvanized coating, AISI 316 rebar, and AISI 304 rebar, which significantly improved the corrosion resistance of reinforced concrete. Compared to noncoated MS reinforcement, the corrosion resistance increased by 2.57, 4.96, and 1.96 times, respectively. Additionally, the inclusion of 2.5 vol.% polypropylene fibers decreased the corrosion rate of steel reinforcement by 2.6 times. After 500 hours of anodic potential application, the polyurethane top coating and zinc-rich epoxy primer exhibited superior adhesion to the concrete surface, with values of 0.87 and 0.60 compared to the control sample. Mathew et al. [33] synthesized the chitosan/boron nitride (CS-BN) nanocomposite using the sol–gel process and incorporated it into an epoxy matrix. FTIR spectrum of the chitosan–boron nitride composite indicates a shift of the B-N bending and stretching vibration peaks to 528 and 1,365 cm^{-1}, respectively. Thermogravimetric analysis (TGA) suggests that the composite exhibits greater thermal stability than pure chitosan, attributed solely to boron nitride. In electrochemical impedance spectroscopy (EIS), the formation of a protective thin film of inhibitor chitosan/boron nitride (CS-BN) composite on the MS surface is evidenced by the increase in net resistance (R_{net}) value as the concentration of boron nitride in the composite increases up to 4%. However, simultaneously, the net capacitance (C_{net}) decreases from 17.1008 $\mu F/cm^2$ to 7.6348 $\times$ 10^{-1} $\mu F/cm^2$. The water contact angles on the surface of MS increased from 54.2° to 63.4° and 77.01° due to composite coating, with an increase in the concentration of boron nitride. The X-ray photoelectron spectroscopy (XPS) spectrum reveals a decrease in the number of corrosion products (Fe^{2+}, Fe^{3+}, and Fe_2O_3) with increased boron nitride composition in the composite. Additionally, the satellite peak of Fe_2O_3 is nearly absent in corroded bare MS coated with CS-BN composite containing 4% BN. Suwasono et al. [34] reported in their study that applying two layers of epoxy coating was effective for anticorrosive purposes for 576 hours. The measured result of the average adhesive power of the epoxy coating was 4,205 MPa. Nguyen et al. [35] studied the adhesion test for coatings, which demonstrated that including nano-Fe_2O_3 and nanoclay particles notably increased the adhesion strength of the epoxy coating. Additionally, the ball impact test revealed a substantial improvement in impact strength, rising from 130 kg·cm to 200 kg·cm with the incorporation of nanoparticles into the epoxy coating. Furthermore, nano-ZnO bolstered the thermal stability of the epoxy, particularly during high-temperature periods. Jeon et al. [36] investigated the reduction in the impedance modulus of log |Z| at 0.01 Hz; during hygrothermal test cycling, it was found to be associated with the initiation of localized blistering on carbon steel coated with epoxy. Conversely, epoxy coatings containing multiwalled carbon nanotubes (MWCNTs) showed no localized blister formation, which was attributed to both enhanced adhesion strength of the epoxy coating and reduced water uptake by the coatings. This improvement aligns closely with enhanced corrosion resistance. Table 8.1 shows the corrosion behavior of steel in marine environment.

Table 8.1 Effect of coating on corrosion behavior of steel

S. No.	Author name	Coating substrate, material, and method	Results
1	Lin et al. [28]	*Substrate:* Stainless steel *Material:* Epoxy resin *Method:* Brushing method	The LEIS impedance reached a stable state faster. As the coating thickness increases, the LEIS impedance shows the corrosion protection of the coating
2	Nooredeen et al. [29]	*Substrate:* Stainless steel 316L *Material:* Epoxy resin *Method:* Paint brush, or roller method	The nano-$CoFe_2O_4$ @SiO_2 forms a passive protective layer at the steel surface, which inhibits the corrosion and enhances the corrosion-protection efficiency of phosphomolybdate (POM)/$CoFe_2O_4$ @SiO_2 nanocomposite coatings, which is confirmed by EIS study
3	Edvardsen et al. [30]	*Substrate:* Stainless steel 316L *Material:* Epoxy and carbon nanofiber (CNF) *Method:* Deposition method	For industrial applications of electroconductive coatings with organic matrix, reducing degradation upon application of pulsed current can be a significant advantage
4	Alam et al. [31]	*Substrate:* Stainless steel *Material:* Epoxy resin bisphenol-A type *Method:* Deposition method	The epoxy coating containing ZrO_2 nanoparticles has better adhesion and slower degradation due to its low porosities. Coating with 2% Zr (EZr-2) shows an excellent corrosion resistance, compared to the 1% and 3% reinforced ZrO_2 nanoparticles
5	Afshar et al. [32]	*Substrate:* MS (ST37), stainless steel (AISI316 and AISI304) *Material:* Polyurethane and zinc-rich epoxy primer *Method:* Hot-dip galvanized coating	Steel reinforcement with hot-dip galvanized coating, AISI 316 rebar, and AISI 304 rebar enhanced the corrosion resistance of reinforced concrete In the first mix design by 2.57, 4.96, and 1.96 times, respectively, over the MS reinforcement that was not coated
6	Mathew et al. [33]	*Substrate:* MS *Material:* Chitosan/boron nitride (CS-BN) incorporated into epoxy resin *Method:* Spin coating method	The composite formation is confirmed by FTIR. BN particles are finely dispersed throughout the chitosan matrix, according to XRD. The TGA results indicate that BN improves the chitosan's thermal characteristics. The production of a thin layer of (CS-BN) composite at the interface between MS and electrolyte is indicated by EIS. The contact angle measurements reveal that the hydrophobicity of polished steel surfaces rises as a result of the composite coating. The composites thin layer creation in the XPS analysis can prevent MS from corroding

(Continued)

Table 8.1 Effect of coating on corrosion behavior of steel (Continued)

S. No.	Author name	Coating substrate, material, and method	Results
7	Suwasono et al. [34]	*Substrate:* Q 235 steel *Material:* Epoxy resins *Method:* Spray method	Average roughness value of coating materials for epoxy was 81.14 μm
8	Nguyen et al. [35]	*Substrate:* CT3 steel *Material:* Epoxy resin *Method:* Dipping method	Epoxy coating significantly enhanced the impact strength from 130 kg·cm to 200 kg·cm
9	Jeon et al. [36]	*Substrate:* Carbon steel (JIS G3131 SPHC) *Material:* Epoxy *Method:* Air spraying method	A decrease in the impedance modulus $|Z|$ in the low-frequency region during the hygrothermal test. The average adhesion strength of the epoxy coating without MWCNTs was 13.58 MPa

8.4 CHARACTERISTICS OF EPOXY RESINS

Epoxy resin belongs to a versatile class of polymer materials distinguished by the presence of oxirane within their molecular structure [37]. Epoxy stands out for its distinctive properties, including low shrinkage, high strength, remarkable environmental degradation resistance, and excellent ability to withstand elevated temperatures. Additionally, epoxy exhibits minimal shrinkage compared to alternative matrices and can be cured at temperatures ranging from 5°C to 150°C [38, 39]. Furthermore, epoxy exhibits the advantage of low viscosity in its uncured state, simplifying processing during fabrication without requiring high-pressure equipment. Moreover, composites suitable for structural applications demanding high strength and low weight often employ epoxy resin reinforced with glass fiber [40, 41]. The chemical formula of epoxy resin 1,3-bis(2,3-epoxypropoxy)-benzene is shown in Figure 8.3 [42].

Epoxy resins are typically categorized into three main types: bisphenol-A, bisphenol-F, and novolac. Bisphenol-A, derived from the reaction of phenol and acetone, undergoes further reaction with epichlorohydrin to produce bisphenol-A diglycidyl ether (DGEBA) [43]. This compound is known for its favorable qualities, such as abrasion, wear, and impact resistance. Bisphenol-F, resulting from the reaction of phenol and formaldehyde, undergoes a subsequent reaction with epichlorohydrin to yield bisphenol-F diglycidyl ether (DGEBF). This compound is recognized for conferring superior chemical resistance compared

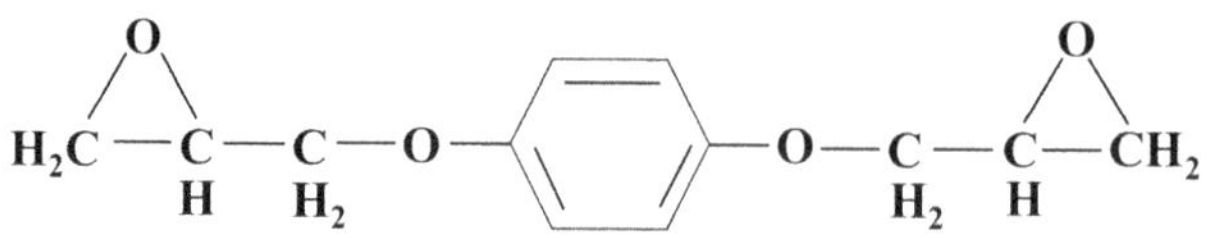

Figure 8.3 Chemical formula of the epoxy resin.

to bisphenol-A. Novolac epoxy resins are derived from modifications of bisphenol-F resins, achieved by reacting an excess of phenol and formaldehyde. This compound provides excellent temperature resistance and is formed by reacting excess phenol and formaldehyde. However, it tends to render the epoxy matrix more brittle than bisphenol-A or bisphenol-F [44].

8.5 UTILIZING NANOFILLERS TO ENHANCE THE PROPERTIES OF EPOXIES

Researchers have recently begun enhancing conventional epoxies by incorporating various micro/nanosized fillers to develop stand-alone or hybrid micro/nanocomposite coatings. This approach addresses the numerous limitations of pristine epoxy systems. Nanofillers have been widely employed to enhance epoxy materials friction, wear resistance, and load-bearing capabilities. Figure 8.4 shows commonly utilized fillers incorporated to modify epoxy resin systems.

8.5.1 Carbon-based filler

Graphite fillers, characterized by their layered structure and weak van der Waals forces between layers, exhibit high electrical conductivity [45]. They also aid in reducing friction and wear by creating a lubricant film between mating surfaces [46]. Zhang et al. [45] reported that incorporating short carbon fiber (SCF) fillers in epoxy matrices is widely utilized in industrial applications because incorporating SCF enhances the compressive strength and creep resistance of epoxy matrices. Graphene and graphene oxide exhibit remarkable properties, including high fracture toughness and mechanical strength, as demonstrated by Young's modulus values of 1 TPa and tensile strength of 130 GPa. They also possess high thermal and electrical conductivities [47–49]. Graphene confers water- and oil-resistant characteristics to the epoxy matrix, whereas graphene

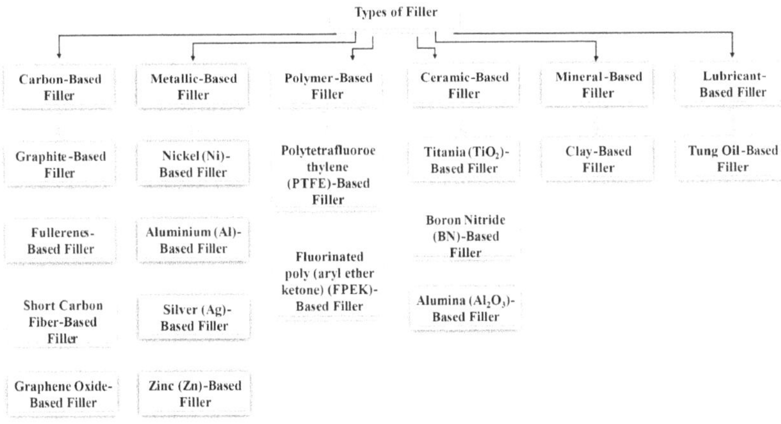

Figure 8.4 Commonly utilized fillers incorporated to modify epoxy resin systems.

oxide is hydrophilic. Both graphene and graphene oxide are recognized for their ability to enhance the glass transition temperature of epoxy polymers. This attribute renders them especially valuable in formulating coating and composite systems with higher temperature resistance [50].

8.5.2 Metallic-based filler

Brostow et al. [51] studied the effect of incorporating metallic fillers on MS substrates. They incorporated the micrometer-sized nickel, aluminum, zinc, and silver powders on the MS substrates. They found that coatings' tribological properties and surface energy characteristics were enhanced compared to uncoated MS. It is also reported that the epoxy coating systems modified with aluminum fillers exhibited the lowest friction and wear during pin-on-disk testing. This was attributed to their spherical morphology, enabling them to function like roller bearings.

8.5.3 Polymer-based filler

Zhang et al. [45] and Brostow et al. [51] studied the effect of polymer-based fillers like fluorinated polyaryletherketone (FPEK) and polytetrafluoroethylene (PTFE), respectively, to enhance the properties of epoxy-based composites. They reported that additive FPEK increases the scratch resistance while using polytetrafluoroethylene fillers increases the coefficient of friction of composite compared to unfilled specimen.

8.5.4 Ceramic-based filler

Ceramics comprise metallic atoms with oxides, carbides, borides, nitrides, silicides, and other elements. They are recognized for their characteristic of hot hardness, which enables them to maintain mechanical properties even at elevated temperatures [52]. Pinto et al. [53] studied the mechanical characteristics of epoxy nanocomposites reinforced with titanium dioxide fillers. They reported that the boron nitride ceramic filler shows 20% higher interfacial adhesion to the epoxy matrix than carbon nanotubes (CNTs). These fillers exhibited remarkable performance in wear and friction tests. The aluminum oxide fillers show a high Young's modulus of approximately 347 GPa, and incorporating these fillers enhances the compressive strength and wear resistance of unmodified epoxy matrices [54, 55].

8.5.5 Mineral-based filler

Another category of nanosized fillers employed in epoxy modification is montmorillonites, often referred to as nanoclays. Nanoclays enhance mechanical strength and wear resistance; however, precise control over filler loading is essential. Higher filler loading can lead to particle agglomeration, uneven dispersion, and undesirable properties. Pristine clay is inherently hydrophilic and does not establish a strong bond with the epoxy matrix [56, 57].

8.5.6 Lubricant-based filler

Epoxy coatings were created by adding polysulfone microcapsules containing tung oil, considering the oil's self-healing and self-lubricating properties. Polysulfone was selected as the encapsulating material because of its exceptional physical, thermal, and chemical stability and mechanical strength. The wear rate, friction coefficient, and corrosion rate all showed a significant decline under the composite coating. This corrosion resistance and self-lubrication improvement was attributed to the transfer film forming [58]. Properties enhancement of epoxy matrices with the addition of nanofillers can be seen in the Table 8.2.

Table 8.2 Nanofillers utilized for modifying epoxy matrices and their predominant properties

Sl. No.	Filler material	Filler group	Properties	Reference
1	Graphite	Carbon-based	The graphite filler increases the wear and friction resistance by adding a lubricant coating facilitated by high electrical conductivity	[45,46]
2	CNTs	Carbon-based	Adding CNTs increases the high electrical and thermal conductivity and excellent mechanical properties such as strength and stiffness. The exceptional mechanical properties of CNTs contribute to enhanced tensile strength and modulus and the flexural strength of coated specimens. Additionally, they reduce wear by functioning as roller bearings between contacting surfaces	[59,60,61]
3	SCF	Carbon-based	Electrical and thermal conductivity, as well as mechanical strength, are provided by carbon layers oriented parallel to the fiber axis. This orientation improves compressive strength and creep resistance	[45]
4	Graphene and graphene oxide	Carbon-based	Graphene boasts high mechanical strength, fracture toughness, and electrical and thermal conductivity, surpassing CNTs in dispersibility. Its integration into the epoxy matrix enhances mechanical properties and wear resistance. Graphene exhibits water- and oil-repellent properties, while graphene oxide is hydrophilic. Both contribute to elevating the coatings glass transition temperature (T_g)	[47,50,62,63, 64,65, 66]
5	Fullerenes (C_{60})	Carbon-based	These additives exhibit remarkably high tensile strength at lower loading than ceramic fillers. Their spherical shape reduces the tendency to agglomerate, making them significantly more straightforward to disperse than nanoclays and CNTs	[67]

(Continued)

Table 8.2 Nanofillers utilized for modifying epoxy matrices and their predominant properties (Continued)

Sl. No.	Filler material	Filler group	Properties	Reference
6	Nickel, aluminum, silver, and zinc	Metallic-based	Aluminum demonstrates the lowest friction and wear due to its spherical shape, which functions as roller bearings. However, zinc or zinc oxide exhibits the highest friction and wear	[51]
7	Polytetrafluoroethylene (PTFE)	Polymer-based	It acts as a lubricant and is one of the conventional fillers capable of reducing the frictional coefficient by forming a transfer film	[45,68]
8	Fluorinated poly aryl ether ketone (FPEK)	Polymer Based	This filler material is recognized for its exceptional hydrophobicity and ability to enhance scratch resistance.	[51]
9	Titania (TiO_2)	Ceramic-based	It enhances the wear and impact resistance, fracture toughness, and coating pull-off strengths. The nanosized TiO_2 particles provide a higher level of wear resistance to the epoxy matrix than micrometer-sized TiO_2 fillers under identical conditions	[45,53,69]
10	Boron nitride (BN)	Ceramic-based	Boron nitride fillers provide high thermal conductivity and enhance mechanical properties, UV radiation resistance, oxidation resistance, and thermal stability. It demonstrates 20% higher interfacial adhesion to the epoxy matrix than CNTs	[70,71]
11	Alumina (Al_2O_3)	Ceramic-based	Alumina enhance compressive strength and wear resistance	[54,55]
12	E-glass fibers	Ceramic-based	Glass fibers are widely used reinforcement for polymer matrices and are renowned for enhancing mechanical properties. Nevertheless, their performance falls notably lower when contrasted with SCF reinforcements	[54]
13	Nanoclay (montmorillonite)	Mineral-based	A highly prevalent class of nanofillers, they are recognized for enhancing mechanical properties and imparting high wear resistance. However, particle agglomeration is a commonly encountered issue with these fillers	[56,57]
14	Tung oil	Lubricant-based	It provides mechanical strength and remarkable physical, thermal, and chemical stability. It decreases the wear rate, friction coefficient, and corrosion rate. Additionally, it showcases self-healing and self-lubricating characteristics	[58]

8.6 NOVEL ORGANIC COATING MATERIALS DESIGNED FOR MARINE APPLICATIONS

Using high-performance anticorrosion coatings is a simple and effective way to address metal corrosion. International and local material research continues to be heavily focused on the hunt for cutting-edge anticorrosion coatings designed explicitly for maritime vessels. Several leading-edge, high-performance anticorrosion coatings are intended for ships below. The epoxy resin molecule contains hydroxyl, ether, and epoxy groups, which are polar structures. These groups exhibit strong adhesive properties to the surface they are applied to, which can also be interdiffused due to the compatibility of their metal oxides. Upon curing, the epoxy resin forms stable benzene rings and ether bonds between its molecules, rendering the resulting film resistant to organic solvents and various chemical agents. In contemporary ship anticorrosion coatings, epoxy resin coatings are predominantly modified, such as with inorganic fillers or tar epoxy resin. These coatings typically fall into high-solid content and water-based formulations [72]. A high-solid coating is a coating system with a solid content exceeding 70% during application. This type of coating exhibits reduced levels of volatile organic compounds (VOCs), resulting in higher coating density post-curing and offering outstanding weathering resistance and mechanical strength. Jiafeng et al. [73] studied the novel, enduring epoxy coating developed for ship splash zones prone to corrosion. The team opted for bisphenol-A liquid epoxy resin and modified resin as the fundamental components of the anticorrosive coating. At the same time, the curing agent was chosen from modified polyamide and phenylamine. The extender level is typically raised in amine-cured epoxy coatings to prevent coating sagging during application. Utilizing low molecular weight in epoxy high-solids coatings makes them more brittle than traditional coatings [74]. The coating system exhibits four key characteristics. First, it demonstrates strong adhesion to diverse substrates. Second, waterborne coatings possess low VOCs, thus minimizing environmental pollution. Third, water-based coatings utilize water as a dispersion medium, making them safer during storage, transportation, and application. Fourth, waterborne coatings feature rapid drying properties. Under conditions of 20°C temperature and 75% relative humidity, surface drying typically occurs within 2 hours [75]. Polyurethane coatings are distinguished by their excellent water resistance, acid solubility, superior alkali resistance, favorable mechanical properties, and impressive weather resistance compared to most polymers. They result from isocyanates (R—N=C=O) reaction with reactive water molecules or unstable hydrogen atoms. The crosslinking process of polyurethane coatings is also influenced by factors such as the type and structure of the isocyanate and curing agent, temperature, and the presence of catalysts. The oligomeric polyisocyanates' attributes of low viscosity, high functionality, and elevated isocyanate content render them especially well-suited for high-solid content coatings in polyurethane applications. Techniques employed in developing high-solid content polyurethane coatings for epoxy coatings involve strategies such as diminishing resin binders, incorporating diluents, or lowering the viscosity of polyisocyanate crosslinking agents [76]. In alignment with the principles of green chemistry, high-solid polyurethane coatings can also be formulated using plant-derived polyols. An innovative UV curable waterborne polyurethane-acrylate (WPUA) was synthesized through sequential emulsion polymerization, utilizing hydroxyl-terminated natural rubber (HTNR) as a primary raw material [77]. Waterborne polyurethanes are a type of binary colloid water-soluble polymer. They are created by adding hydrophilic groups to the polyurethane molecular chains, such as carboxylate, sulfonate, quaternary ammonium salt, or hydrophilic segments [78]. This alteration gives the polyurethane hydrophilic qualities. Paints made of waterborne polyurethane use water as the dispersion medium and have minimal organic solvent. Their attributes include nontoxicity, environment-friendliness, flame resistance, and energy efficiency [79]. To improve the fluoride-containing waterborne polyurethane's film-forming ability and mechanical strength, aziridine crosslinking was used. WFPU's exceptional mechanical and film-forming properties make them appropriate for use as waterproof paints. Waterborne polyurethanes are widely used in many industries, such as paints for wooden furniture, architectural coatings, metal anticorrosion coatings, and many more decorative coating applications [80]. Graphene possesses a two-dimensional lamellar structure with tightly bonded carbon–carbon bonds on its planar layer. This structural configuration imparts exceptional stability to graphite and endows it with outstanding mechanics, electricity, optics, and heat properties. When incorporated into coatings, graphene facilitates the formation of a dense isolation layer through barrier stacking via physical methods. This layer effectively impedes the passage of corrosive media. Graphene is an excellent shielding agent, effectively segregating the metal matrix from the surrounding environment. Waterborne polyurethane, also referred to as water-dispersible polyurethane, water polyurethane, or waterborne polyurethane, is a novel polyurethane formulation that utilizes water as a suspending agent instead of organic solvents. This innovation offers several advantages, including zero pollution, reliability, high physical performance, excellent solubility, and ease of material modification. In the development of waterborne polyurethane anticorrosive coatings, inorganic nanofillers such as graphene oxide (GO), reduced graphene oxide (RGO), and functionalized graphene derivatives are added to enhance performance [81]. Sodium polyacrylate is utilized to achieve uniform and stable dispersion of graphene suspension in aqueous solutions. Upon graphene incorporation, the composite layer exhibits enhanced watertight barrier properties, with a notable reduction in the diffusion rate of water molecules within the layer, leading to significantly improved corrosion resistance. Furthermore, the density of the composite ethylene base is substantially reduced, while the coating resistance and load transfer resistance are increased [82]. A functionalized silane bridging agent (2-(3,4-epoxycyclohexyl)ethyltrioxysilane) was synthesized to enhance the surface modification of graphene oxide, aimed at mitigating the agglomeration of nanomaterials and enhancing the coatings performance through functionalization. This approach was undertaken to evaluate the corrosion resistance of epoxy coatings filled with graphene oxide [83].

8.7 APPLICATIONS OF EPOXY COATINGS

Epoxy resins are widely used because of their exceptional qualities as strong anticorrosion coatings. These include excellent adhesion to various substrates, toughness, low shrinkage upon curing, remarkable solvent and chemical resistance, ease of processing, high safety, and excellent adhesion to solvents and chemicals. They frequently prevent corrosion in metal cans and containers, especially those used to package acidic foods. High-performance and ornamental flooring applications are also useful for epoxy resins [84, 85]. Figure 8.5 shows the paint and coatings applied to protect valves.

In the electronics industry, epoxy resin formulations are essential since they are used in many products, including bushings, insulators, transformers, motors, generators, and switchgear. These electronic parts are also required to protect the marine environment. Epoxy resins are well-known for providing outstanding electrical insulation protecting electrical components from moisture, dust, and short circuits. Electronic gadgets that shield against electromagnetic interference often use metal-filled polymers [86]. The underwater hull and boot top, ballast tanks, topside and superstructure, and deck sides are always in a sea environment that requires protection against corrosion. Coating systems designed for the underwater components of a ship require properties such as corrosion inhibition, antifouling capabilities, abrasion resistance, smoothness, and compatibility with cathodic protection. These systems often use high-solid and occasionally high-build materials to enhance their effectiveness.

Figure 8.5 Images of paints and coatings on valves.

The utilization of epoxy mastics is becoming more prevalent in this regard. In addition to exhibiting good surface tolerance, these coatings typically offer extended maximum overcoat times and excellent recoatability [87]. Figure 8.6 shows the shell plating or hull of the ship undergoes painting each time it enters a dry dock period. The painting has been done to safeguard the metal against various corrosive effects and to uphold its overall. The painted section remains consistently submerged in seawater while the ship is underway. Ballast tank coating systems need to be impervious to cathodic protection, pore-free, and able to tolerate corrosion and the corrosive impacts of seawater. Poor coating quality, mechanical damage, and uneven coating can readily lead to pitting corrosion in ballast tanks. The presence of materials more noble than steel and nearby stainless steel bulkheads from cargo tanks frequently aggravate this corrosion. Sacrificial anodes are occasionally placed next to the coating system to reduce pitting corrosion. A minimum of two layers of high-solids coal tar epoxy, straight epoxy, or modified epoxy with a total dry film thickness (DFT) of at least 250 μm are usually used in modern ballast tank systems [87]. A visually appealing topcoat option could be aliphatic polyurethane or a blend of aliphatic polyurethane/acrylic for the topside and superstructure. Isocyanate-free alternatives, such as epoxy/acrylic or other modified epoxy coatings, are available, although they often exhibit lower gloss and color retention than polyurethanes. Polysiloxane epoxy hybrid coatings can also serve as aesthetic topcoats [87]. Deck paint systems need to be highly resistant to weathering and corrosion. They should also resist impact, scratches, and abrasion and have non-slip qualities, especially in damp environments. They must also be resistant to various contaminants, such as cleaning agents, fuel oils, lubricating greases, seawater, and cargo spills. It is essential to apply a recoatable epoxy holding primer during construction and finish the final coating system as soon as feasible before delivery to minimize any damage to deck coatings [87].

Figure 8.6 The shell plating or hull of the ship undergoes painting to safeguard the metal against various corrosive effects, and the inset displays the hull before painting [88].

8.8 CONCLUSION

This chapter presents a comprehensive review of epoxy resin-based corrosion-resistant coatings for marine engineering applications. It covers the introduction of corrosion in ship structures and its prevention, principles and characteristics of epoxy resins, and the utilization of nanofillers to enhance the properties of epoxies. Recent research highlights the growing interest in developing epoxy resin composites with multiple anticorrosion strategies, which are deemed more cost-effective and efficient than single-strategy approaches. Incorporating nanoparticles into epoxy formulations have enhanced the performance of coatings across various applications, including protective coatings in marine environments. These coatings surface and bulk properties significantly influence their performance, as evidenced by extensive literature reports. The widespread applications of epoxy resin-based coatings in industries such as paints, coatings, electronics, and various shipbuilding areas indicate a promising future. Expectations include a substantial increase in environment-friendly epoxy resin coatings and intensified research efforts from the scientific community in the coming years.

ACKNOWLEDGMENTS

School of Marine Engineering and Technology, Indian Maritime University, Kolkata Campus and Ministry of Ports, Shipping and Waterways. The government of India provided the necessary infrastructure for the authors to conclude this research.

DECLARATION

The authors declare that they have no known competing financial interests or personal relationships that could have appeared to influence the work reported in this chapter.

CONFLICT OF INTEREST

The author(s) declared no potential conflicts of interest with respect to the research, authorship, and/or publication of this chapter.

FUNDING

The author received no financial support for this chapter's research, authorship, and/or publication.

AUTHOR'S CONTRIBUTIONS

Authors collectively contribute to the chapter's design, framing, and writing.

REFERENCES

1. Sangaj, N. S., & Malshe, V. C. (2004). Permeability of polymers in protective organic coatings. *Progress in Organic Coatings*, *50*(1), 28–39.
2. Templeton, D. M., & Liu, Y. (2010). Multiple roles of cadmium in cell death and survival. *Chemico-Biological Interactions*, *188*(2), 267–275.
3. Gardiner, C. P., & Melchers, R. E. (2001). Enclosed atmospheric corrosion in ship spaces. *British Corrosion Journal*, *36*(4), 272–276.
4. Alhumade, H., Abdala, A., Yu, A., Elkamel, A., & Simon, L. (2016). Corrosion inhibition of copper in sodium chloride solution using polyetherimide/graphene composites. *The Canadian Journal of Chemical Engineering*, *94*(5), 896–904.
5. Shelly, D., Singh, K., Nanda, T., & Mehta, R. (2018). Addition of nanomer clays to GFRPs for enhanced impact strength and fracture toughness. *Materials Research Express*, *5*(10), 105013.
6. May, C. (Ed.). (2018). *Epoxy Resins: Chemistry and Technology*. Routledge.
7. Karami, Z., Jazani, O. M., Navarchian, A. H., & Saeb, M. R. (2018). State of cure in silicone/clay nanocomposite coatings: The puzzle and the solution. *Progress in Organic Coatings*, *125*, 222–233.
8. Jouyandeh, M., Paran, S. M. R., Shabanian, M., Ghiyasi, S., Vahabi, H., Badawi, M., ..., & Saeb, M. R. (2018). Curing behavior of epoxy/Fe_3O_4 nanocomposites: A comparison between the effects of bare Fe_3O_4, Fe_3O_4/SiO_2/chitosan and Fe_3O_4/SiO_2/chitosan/imide/phenylalanine-modified nanofillers. *Progress in Organic Coatings*, *123*, 10–19.
9. Wernik, J. M., & Meguid, S. A. (2014). On the mechanical characterization of carbon nanotube reinforced epoxy adhesives. *Materials & Design*, *59*, 19–32.
10. Sancaktar, E., & Kuznicki, J. (2011). Nanocomposite adhesives: Mechanical behavior with nanoclay. *International Journal of Adhesion and Adhesives*, *31*(5), 286–300.
11. Aradhana, R., Mohanty, S., & Nayak, S. K. (2018). High performance epoxy nanocomposite adhesive: Effect of nanofillers on adhesive strength, curing and degradation kinetics. *International Journal of Adhesion and Adhesives*, *84*, 238–249.
12. Azeez, A. A., Rhee, K. Y., Park, S. J., & Hui, D. (2013). Epoxy clay nanocomposites—processing, properties and applications: A review. *Composites Part B: Engineering*, *45*(1), 308–320.
13. Shi, X., Nguyen, T. A., Suo, Z., Liu, Y., & Avci, R. (2009). Effect of nanoparticles on the anti-corrosion and mechanical properties of epoxy coating. *Surface and Coatings Technology*, *204*(3), 237–245.
14. Bagherzadeh, M. R., & Mahdavi, F. (2007). Preparation of epoxy–clay nanocomposite and investigation on its anticorrosive behavior in epoxy coating. *Progress in Organic Coatings*, *60*(2), 117–120.
15. Ramezanzadeh, B., Niroumandrad, S., Ahmadi, A., Mahdavian, M., & Moghadam, M. M. (2016). Enhancement of barrier and corrosion protection performance of an epoxy coating through wet transfer of amino functionalized graphene oxide. *Corrosion Science*, *103*, 283–304.
16. Thakur, V. K., & Kessler, M. R. (2015). Self-healing polymer nanocomposite materials: A review. *Polymer*, *69*, 369–383.

17. Qing, Y., Zhou, W., Luo, F., & Zhu, D. (2010). Epoxy-silicone filled with multi-walled carbon nanotubes and carbonyl iron particles as a microwave absorber. *Carbon, 48*(14), 4074–4080.
18. Camino, G., Tartaglione, G., Frache, A., Manferti, C., & Costa, G. (2005). Thermal and combustion behaviour of layered silicate–epoxy nanocomposites. *Polymer Degradation and Stability, 90*(2), 354–362.
19. Gu, H., Ma, C., Gu, J., Guo, J., Yan, X., Huang, J., ..., & Guo, Z. (2016). An overview of multifunctional epoxy nanocomposites. *Journal of Materials Chemistry C, 4*(25), 5890–5906.
20. Jiang, T., Kuila, T., Kim, N. H., Ku, B. C., & Lee, J. H. (2013). Enhanced mechanical properties of silanized silica nanoparticle attached graphene oxide/epoxy composites. *Composites Science and Technology, 79*, 115–125.
21. Ahmadi, Z. (2019). Epoxy in nanotechnology: A short review. *Progress in Organic Coatings, 132*, 445–448.
22. Mostafaei, A., & Nasirpouri, F. (2013). Preparation and characterization of a novel conducting nanocomposite blended with epoxy coating for antifouling and antibacterial applications. *Journal of Coatings Technology and Research, 10*, 679–694.
23. Harsimran, S., Santosh, K., & Rakesh, K. (2021). Overview of corrosion and its control: A critical review. *Proceedings of Engineering Sciences, 3*(1), 13–24.
24. Victoria Aleksandrovna Plotkina (2021). Reliability performance of shipboard power complexes with regard to ecosystem effects. *E3S Web of Conferences 320*, 01005.
25. Manjunatha, H., Ratnam, K. V., Janardan, S., Nadh, R. V., Kumar, N. S., Prasad, N. K., ..., & Naidu, K. C. B. (2021). Marine corrosion. *Corrosion Science: Modern Trends and Applications, 174*.
26. Shaw, B., & Kelly, R. (2006). What is corrosion? *The Electrochemical Society Interface, 15*(1), 24.
27. Bleile, H., & Rodgers, S. D. (2001). *Marine Coatings* (pp. 5174–5185).
28. Lin, Y., Li, Z., Wang, X., Liu, X., Chi, J., & Zhang, Z. (2024). Effect of coating damage on the micro area corrosion performance of HDR duplex stainless steel. *Coatings, 14*(2), 174.
29. Nooredeen, N. M., Youssef, E. A., Mousa, A. R. M., El-Ghaffar, A., & Ahmed, M. (2023). Study of CeO_2@ TiO_2/POM and $CoFe_2O_4$@ SiO_2/POM composites as highly efficient eco-friendly anti-corrosion coating for 316 stainless steel. *Egyptian Journal of Chemistry, 66*(3), 37–52.
30. Edvardsen, L., Grandcolas, M., Lædre, S., Yang, J., Lange, T., Bjørge, R., & Gaweł, K. (2024). The role of electrical current mode in calcium carbonate deposition on conductive carbon nanofiber–epoxy coating. *Journal of Applied Polymer Science, 141*(13), e55152.
31. Alam, M. A., Samad, U. A., Anis, A., Sherif, E. S. M., Abdo, H. S., & Al-Zahrani, S. M. (2023). The effect of zirconia nanoparticles on thermal, mechanical, and corrosion behavior of nanocomposite epoxy coatings on steel substrates. *Materials, 16*(13), 4813.
32. Afshar, A., Jahandari, S., Rasekh, H., Rahmani, A., & Saberian, M. (2023). Effects of different coatings, primers, and additives on corrosion of steel rebars. *Polymers, 15*(6), 1422.
33. Mathew, Z. P., Shamnamol, G. K., Greeshma, K. P., & John, S. (2023). Insight on the corrosion inhibition of nanocomposite chitosan/boron nitride integrated epoxy coating system against mild steel. *Corrosion Communications, 9*, 36–43.
34. Suwasono, B., Putra, I. K. A. S., Kristiyono, T. A., & Azhar, A. (2021). Adhesive coating value based on the main ingredient of ship paint. *Brodogradnja: Teorija i praksa brodogradnje i pomorske tehnike, 72*(2), 19–36.
35. Nguyen, T. A., Nguyen, T. V., Thai, H., & Shi, X. (2016). Effect of nanoparticles on the thermal and mechanical properties of epoxy coatings. *Journal of Nanoscience and Nanotechnology, 16*(9), 9874–9881.

36. Jeon, H., Park, J., & Shon, M. (2013). Corrosion protection by epoxy coating containing multi-walled carbon nanotubes. *Journal of Industrial and Engineering Chemistry, 19*(3), 849–853.

37. Broomfield, L. M., Sebastián, R. M., Marquet, J., & Schönfeld, R. (2012). Ambient temperature polymerization of oxiranes initiated by the novel MSbF6/H$_2$O co-initiator system. *Polymer, 53*(25), 5632–5640.

38. Kandpal, J., B Yadaw, S., & K Nagpal, A. (2013). Mechanical properties of multifunctional epoxy resin/glass fiber reinforced composites modified with poly(ether imide). *Advanced Materials Letters, 4*(3), 241–249.

39. Ratna, D. (2007). *Epoxy Composites: Impact Resistance and Flame Retardancy* (Vol. 16). iSmithers Rapra Publishing.

40. Jaya Vinse Ruban, Y., Ginil Mon, S., & Vetha Roy, D. (2013). Mechanical and thermal studies of unsaturated polyester-toughened epoxy composites filled with amine-functionalized nanosilica. *Applied Nanoscience, 3*, 7–12.

41. Pihtili, H. (2009). An experimental investigation of wear of glass fibre–epoxy resin and glass fibre–polyester resin composite materials. *European Polymer Journal, 45*(1), 149–154.

42. Iwamoto, N., Yuen, M. M., & Fan, H. (2012). *Molecular Modeling and Multiscaling Issues for Electronic Material Applications* (Vol. 57). A. Wymysłowski (Ed.). London: Springer.

43. Pham, H. Q., & Marks, M. J. (2000). Epoxy resins. Ullmann's Encyclopedia of Industrial Chemistry. Weinheim, Germany: Wiley-VCH.

44. Bobby, S., & Samad, M. A. (2017). Enhancement of tribological performance of epoxy bulk composites and composite coatings using micro/nano fillers: A review. *Polymers for Advanced Technologies, 28*(6), 633–644.

45. Zhang, Z., Breidt, C., Chang, L., Haupert, F., & Friedrich, K. (2004). Enhancement of the wear resistance of epoxy: Short carbon fibre, graphite, PTFE and nano-TiO2. *Composites Part A: Applied Science and Manufacturing, 35*(12), 1385–1392.

46. Pan, G., Guo, Q., Ding, J., Zhang, W., & Wang, X. (2010). Tribological behaviors of graphite/epoxy two-phase composite coatings. *Tribology International, 43*(8), 1318–1325.

47. Shah, R., Datashvili, T., Cai, T., Wahrmund, J., Menard, B., Menard, K. P., ..., & Perez, J. (2015). Effects of functionalized reduced graphene oxide on frictional and wear properties of epoxy resin. *Materials Research Innovations, 19*(2), 97–106.

48. Svendsen, E. M. (2014). *Graphene Oxide as Reinforcement in Epoxy Based Nanocomposites* (Master's thesis, NTNU).

49. Kumar, A. M. (2016). Hybrid nanocomposite from graphene oxide and TiO$_2$ nanoparticles: Surface protective performance on steel substrates. In *16th Middle East Corrosion Conference and Exhibition.*

50. Tong, Y., Bohm, S., & Song, M. (2013). Graphene based materials and their composites as coatings. *Austin Journal of Nanomedicine & Nanotechnology, 1*(1), 1003–1019.

51. Brostow, W., Dutta, M., & Rusek, P. (2010). Modified epoxy coatings on mild steel: Tribology and surface energy. *European Polymer Journal, 46*(11), 2181–2189.

52. Subhash, N. V. S. S. (2013). *Tribology of Alumina Nano Composites* (Doctoral dissertation).

53. Pinto, D., Bernardo, L., Amaro, A., & Lopes, S. (2015). Mechanical properties of epoxy nanocomposites using titanium dioxide as reinforcement: A review. *Construction and Building Materials, 95*, 506–524.

54. Mohanty, A., & Srivastava, V. K. (2015). Effect of alumina nanoparticles on the enhancement of impact and flexural properties of the short glass/carbon fiber reinforced epoxy-based composites. *Fibers and Polymers, 16*, 188–195.

55. Jia, Q. M., Zheng, M., Xu, C. Z., & Chen, H. X. (2006). The mechanical properties and tribological behavior of epoxy resin composites modified by different shape nanofillers. *Polymers for Advanced Technologies, 17*(3), 168–173.
56. Wang, L., Wang, K., Chen, L., Zhang, Y., & He, C. (2006). Preparation, morphology and thermal/mechanical properties of epoxy/nanoclay composite. *Composites Part A: Applied Science and Manufacturing, 37*(11), 1890–1896.
57. Esteves, M., Ramalho, A., Ferreira, J. A. M., & Nobre, J. P. (2013). Tribological and mechanical behaviour of epoxy/nanoclay composites. *Tribology Letters, 52*, 1–10.
58. Li, H., Cui, Y., Wang, H., Zhu, Y., & Wang, B. (2017). Preparation and application of polysulfone microcapsules containing tung oil in self-healing and self-lubricating epoxy coating. *Colloids and Surfaces A: Physicochemical and Engineering Aspects, 518*, 181–187.
59. Garg, M., Sharma, S., & Mehta, R. (2015). Pristine and amino functionalized carbon nanotubes reinforced glass fiber epoxy composites. *Composites Part A: Applied Science and Manufacturing, 76*, 92–101.
60. Zhou, Y. X., Wu, P. X., Cheng, Z. Y., Ingram, J., & Jeelani, S. (2008). Improvement in electrical, thermal and mechanical properties of epoxy by filling carbon nanotube. *Express Polymer Letters, 2*(1), 40–48.
61. Khun, N. W., Zhang, H., Yang, J., & Liu, E. (2013). Mechanical and tribological properties of epoxy matrix composites modified with microencapsulated mixture of wax lubricant and multi-walled carbon nanotubes. *Friction, 1*, 341–349.
62. Khun, N. W., Zhang, H., Lim, L. H., & Yang, J. (2015). Mechanical and tribological properties of graphene modified epoxy composites. *Applied Science and Engineering Progress, 8*(2), 101–109.
63. Galpaya, D., Wang, M., George, G., Motta, N., Waclawik, E., & Yan, C. (2014). Preparation of graphene oxide/epoxy nanocomposites with significantly improved mechanical properties. *Journal of Applied Physics, 116*(5), 053518.
64. Wajid, A. S., Ahmed, H. T., Das, S., Irin, F., Jankowski, A. F., & Green, M. J. (2013). High-performance pristine graphene/epoxy composites with enhanced mechanical and electrical properties. *Macromolecular Materials and Engineering, 298*(3), 339–347.
65. Abdullah, S. I., & Ansari, M. N. M. (2015). Mechanical properties of graphene oxide (GO)/epoxy composites. *HBRC Journal, 11*(2), 151–156.
66. Daloia, D. (2014). *Friction and Wear Behavior of Graphene Reinforced Epoxy* (Master's thesis, University of Dayton).
67. Rafiee, M. A., Yavari, F., Rafiee, J., & Koratkar, N. (2011). Fullerene–epoxy nanocomposites-enhanced mechanical properties at low nanofiller loading. *Journal of Nanoparticle Research, 13*, 733–737.
68. McCook, N. L., Boesl, B., Burris, D. L., & Sawyer, W. G. (2006). Epoxy, ZnO, and PTFE nanocomposite: Friction and wear optimization. *Tribology Letters, 22*, 253–257.
69. Bhagat, S. (2010). Effect of filler parameter on microstructure and mechanical properties of titanium dioxide reinforced epoxy composites. *Journal of Environmental Nanotechnology, 2*(2), 29–33.
70. Düzcükoğlu, H., Ekinci, Ş., Şahin, Ö. S., Avci, A., Ekrem, M., & Ünaldi, M. (2015). Enhancement of wear and friction characteristics of epoxy resin by multiwalled carbon nanotube and boron nitride nanoparticles. *Tribology Transactions, 58*(4), 635–642.
71. Gu, J., Zhang, Q., Dang, J., & Xie, C. (2012). Thermal conductivity epoxy resin composites filled with boron nitride. *Polymers for Advanced Technologies, 23*(6), 1025–1028.
72. Liu, X., Shen, T., & Cheng, H. (2021). Research progress of epoxy resin anticorrosive coatings. *Plastics Science Technology, 49*(9), 96–100.

73. Jiafeng, L., Weiqiang, Z., & Jianjun, F. (2020) Development of high solid content epoxy coating for maintenance in splash zone. *Paint & Coatings Industry*, *50* (4), 58–64.

74. Heaner, W. L., Aguirre-Vargas, F., Ge, S., Xu, N., Krishnan Karunakaran, K., & Weishuhn, J. (2016). Tunning toughness and flexibility in liquid applied high solids epoxy coatings. *Corpus ID*, 102500379.

75. Yingmao, W., Yahao, Z., Congshu, H., Jingjing, W., & Yu, L. (2021). Application and research progress of waterborne anti-corrosion coatings. *Development and Application of Materials*, *36*(6), 91–96.

76. Pélissier, K., & Thierry, D. (2020). Powder and high-solid coatings as anticorrosive solutions for marine and offshore applications? A review. *Coatings*, *10*(10), 916.

77. Tsupphayakorn-aek, P., Suwan, A., Tulyapitak, T., Saetung, N., & Saetung, A. (2022). A novel UV-curable waterborne polyurethane-acrylate coating based on green polyol from hydroxyl telechelic natural rubber. *Progress in Organic Coatings*, *163*, 106585.

78. Liu, X., Hong, W., & Chen, X. (2020). Continuous production of waterborne polyurethanes: A review. *Polymers*, *12*(12), 2875.

79. Scrinzi, E., Rossi, S., Deflorian, F., & Zanella, C. (2011). Evaluation of aesthetic durability of waterborne polyurethane coatings applied on wood for interior applications. *Progress in Organic Coatings*, *72*(1-2), 81–87.

80. Zhao, J., Zhou, T., Zhang, J., Chen, H., Yuan, C., Zhang, W., & Zhang, A. (2014). Synthesis of a waterborne polyurethane-fluorinated emulsion and its hydrophobic properties of coating films. *Industrial & Engineering Chemistry Research*, *53*(49), 19257–19264.

81. Li, J., Cui, J., Yang, J., Li, Y., Qiu, H., & Yang, J. (2016). Reinforcement of graphene and its derivatives on the anticorrosive properties of waterborne polyurethane coatings. *Composites Science and Technology*, *129*, 30–37.

82. Yuqiong, W., Shuan, L., Zhaoping, L., et al. (2015) Waterproof and protective properties of graphene doped waterborne epoxy resin. *Electroplating & Finishing*, *34* (6): 314–319+361.

83. Zhang, C., Dai, X., Wang, Y., Sun, G., Li, P., Qu, L., ..., & Dou, Y. (2019). Preparation and corrosion resistance of ETEO modified graphene oxide/epoxy resin coating. *Coatings*, *9*(1), 46.

84. Katariya, M. N., Jana, A. K., & Parikh, P. A. (2013). Corrosion inhibition effectiveness of zeolite ZSM-5 coating on mild steel against various organic acids and its antimicrobial activity. *Journal of Industrial and Engineering Chemistry*, *19*(1), 286–291.

85. Gergely, A., Bertóti, I., Török, T., Pfeifer, É., & Kálmán, E. (2013). Corrosion protection with zinc-rich epoxy paint coatings embedded with various amounts of highly dispersed polypyrrole-deposited alumina monohydrate particles. *Progress in Organic Coatings*, *76*(1), 17–32.

86. Chand, N., & Nigrawal, A. (2008). Development and electrical conductivity behavior of copper-powder-filled-epoxy graded composites. *Journal of Applied Polymer Science*, *109*(4), 2384–2387.

87. Berendsen, A. M. (1998). Ship Painting: Current Practice and Systems in Europe (pp. 24–33). Proceedings of Protective Coating Europe.

88. Furlan, P. Y., Jaravata, E. J., Furlan, A. Y., & Kahl, P. (2023). Will it rust? A set of simple demonstrations illustrating iron corrosion prevention strategies at sea. *Journal of Chemical Education*, *100*(2), 1081–1088.

Chapter 9

Ceramic matrix composites: Advanced manufacturing processes and challenges

Sankata Tiwari, Gagan Bansal, and Santosh Kumar

9.1 INTRODUCTION

Composite materials are gaining interest in many industrial applications because of their excellent strength-to-weight ratio compared to traditional materials. Composites are often categorized as polymer matrix composites (PMCs), metal matrix composites (MMCs), and ceramic matrix composites (CMCs) based on the material that forms the matrix. Figure 9.1 shows the primary composite groups, their various reinforcement designs (particulate or fiber reinforcements), and commonly used materials. Carbide-based ceramics are utilized for applications at very high temperatures (above 2,000°C). These ceramics are often strengthened by incorporating carbon fibers and other high-temperature fibers. Carbon matrix composites are used for short exposure to high-temperature applications. Oxide-based ceramics such as alumina, zirconia, silica, and titania are used where superior high-temperature strength and stability are required. For ultrahigh temperature (UHT) applications, ZrB_2 and HfB_2 matrix composites are used to fabricate randoms, nozzles, and other high-temperature components [1].

Ceramics are extremely desirable for high-temperature applications [2] due to their high hardness, high compressive strength, chemically inert, and excellent high-temperature-resistant properties. Despite having excellent properties, the use of ceramics is limited because of their low fracture toughness [3]. Incorporating several reinforcements, such as particles and fibers inside the ceramic matrix, improves the toughness of ceramics [4]. The unusual combination of features in these new materials is causing a revolution in aerospace, energy, and other strategic sectors. These materials are a unique category of composites in which both the reinforcements and the binding matrix are composed of ceramic elements. This combination produces a substance that surpasses the constraints of CMCs and has a wide range of uses, including application engine components that can endure extremely high temperatures and create lightweight armor for enhanced protection in defense sectors. Including fine particles and fibers in the ceramic matrix arrests the crack propagation, and hence the fracture toughness improves significantly. Figure 9.2 shows the increasing trend in demand of CMCs in various strategic sectors.

DOI: 10.1201/9781003564355-9

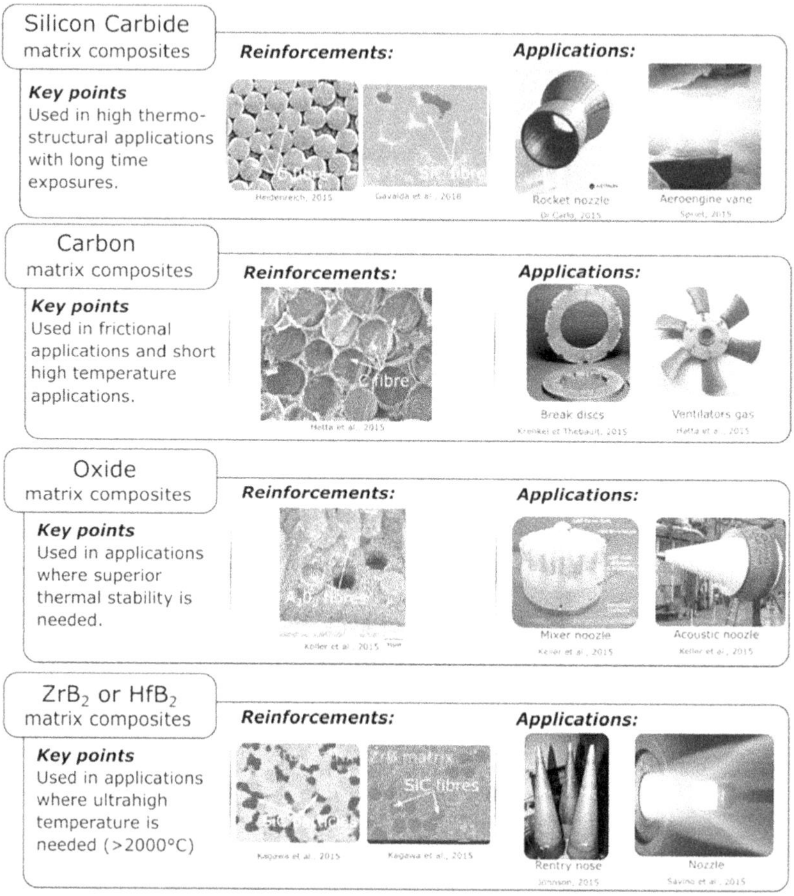

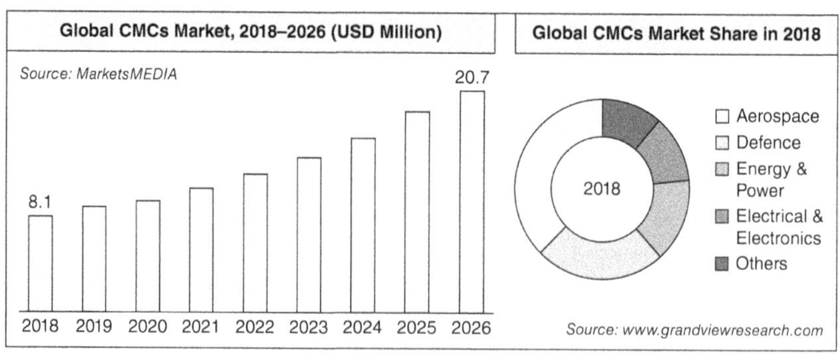

Figure 9.1 Summary presenting the CMC matrices with various reinforcements and their applications in strategic areas. (Reprinted from Diaz et al. [2019]. Copyright 2024, with permission from Elsevier.)

Figure 9.2 Global CMC market between 2018 and 2026 (Reprinted from Sun et al. [2023]. Copyright 2024, with permission from Elsevier.)

As research progresses, it is anticipated that CMCs will assume a more significant role in advancing high-performance machines and structures. Ceramics and ceramics-based composites are also appreciated in orthopedics and dental applications [5]. Conventional ceramics are prone to brittleness and fracture. By integrating fibers, CMCs enhance their toughness and resistance to these failures. Combining strength and lightness envisions a substance that amalgamates the remarkable durability and thermal resilience of ceramics with the lightweight characteristics of composites. Their enhancing properties characterize the phenomenon of CMCs. These innovative materials are transforming industries such as aerospace and energy due to their distinctive combination of features.

Several CMC systems have been fabricated utilizing different synthesis methods to meet specific application needs. Sarkar et al. [6] produced Al_2O_3–ZrO_2 nanocomposite powder using sol–gel synthesis. The synthesized material has been discovered to be useful in tools used for cutting applications and also as component for prostheses since it exhibits unique features, including excellent toughness, wear resistance, and strength at room temperature. Beitollahi et al. [7] examined using sucrose to produce and analyze Al_2O_3–ZrO_2 nanocomposite powder. This procedure involves an amended chemical method in which a powder is created using sucrose, polyvinyl alcohol (PVA), and nitrate salts of various metals. A post-calcination process then follows this. CMCs are creating interest increasingly in researchers as well as in industry because they provide superior thermal and mechanical qualities compared to single-phase ceramic materials. Employing ceramic material as both the matrix and reinforcement has improved several attributes, such as heat conductivity, mechanical strength, and hardness. The characteristics of CMCs are affected by the processing procedure. This study presents numerous processing methods for synthesizing different CMCs. The alumina content ranges from 3% to 48%. In their inquiry, they employed the aqueous combustion synthesis method to get nanopowders. Several researchers have incorporated secondary phases inside the primary ceramic matrix, improving mechanical and physical properties. Fiber and whisker incorporation as reinforcement provides more strength than particulate reinforcement. Toughening of advanced ceramics such as alumina has been achieved by reinforcement of with zirconia, leading to an entirely new material; i.e. zirconia toughened alumina which shows fracture toughness significantly higher with respect to alumina [8].

As previously stated, there are several methods for producing CMCs. The methods encompass the solid-state processing, sol–gel method, plasma spraying technique, laser synthesis method, hydrothermal synthesis, and co-precipitation method. Co-precipitation and hydrothermal approaches are innovative methods for producing nanoparticles and achieving a uniform microstructure in composite materials. Additionally, progress has been made by adapting polymer as a surfactant in the synthesizing process. The different processing methods for the production of CMCs have been illustrated in the following sections.

9.1.1 Processing techniques

Several synthesis processes exist for various CMCs. The different properties such as structural and mechanical properties of CMCs rely on the processing methodology used. Therefore, selecting the processing route is key to achieving the desired properties. Researchers have reported many conventional and advanced processing routes for fabricating different CMCs [9].

9.1.1.1 Sol–gel processing

In sol–gel process, different precursor solutions are combined in a stoichiometric ratio. This solution is maintained at a constant temperature and stirred continuously at room temperature. Ammonium hydroxide solution is incrementally added to this agitated solution. The viscosity of the fluid steadily rises due to the production of hydrogel. The pH has been adjusted to a range of 7–8 in a solution with a suitably more viscosity. The pH plays a major role in synthesis and processing of materials [10–12]. The obtained gel is stored at room temperature for several hours to undergo aging. The formed residue is rinsed away using distilled water. The purpose of washing is to eliminate any leftover ions from the solution. The filtered residue is then dried at 80°C for a few hours. The dehydrated gel is subjected to calcination at an elevated temperature for several hours. The resulting powder is further analyzed using various characterization procedures [13].

9.1.1.2 Laser-based synthesis

Laser synthesis of CMCs involves generating vapor/plasma conditions on the surface by applying strong laser pulses. In this process, a mixture of powders containing various oxides is compressed into pellet form using a certain amount of pressure. The utilization of continuous-wave laser light achieves the laser treatment of this compact. Their creation is linked to elevated temperatures, leading to sintering than melting and solidification [14].

9.1.1.3 Co-precipitation route

In this technique, several metal salts are uniformly combined and made to precipitate using a basic media, often utilizing an ammonia solution. The prepared precipitate is subsequently rinsed to eliminate excess acid and basic ions. The washed product is subjected to drying, calcination, and grinding to get a powder. The utilization of two distinct pathways can achieve co-precipitation. The first method is the direct precipitation technique, where the component to be precipitated is introduced and thoroughly mixed with the salt solution. The second method is the backward/inverse precipitation technique, where the salt solution is mixed and swirled to get the precipitate. In the second procedure, the precipitate is more finely distributed than the precipitate created first. The co-precipitation approach has several benefits. The characteristics of the procedure include simplicity and speed in preparation. It can quickly manipulate particle size and composition [8].

9.2 FIBER-REINFORCED COMPOSITES FABRICATION

These CMCs are produced from a meticulous arrangement of ceramic fibers and a ceramic matrix.

9.2.1 Fiber and Matrix Preparation

- *Fiber processing:* The journey often begins with precursor materials like organic polymers that are converted into high-strength ceramic fibers through pyrolysis (heating in an inert atmosphere).
- *Matrix powder preparation:* Ceramic powders with a desired composition are synthesized through sol–gel processing, where a liquid precursor is transformed into a suspension/gel and then converted into powder.

9.2.2 Fiber Preform Fabrication

- The ceramic fibers are carefully arranged into a desired shape, often through weaving, braiding, or stitching techniques, creating a preform that defines the final CMC component's geometry.

9.2.3 Matrix Infiltration

This crucial step involves filling the gaps within the fiber preform with the ceramic matrix material. There are two main approaches:

- *Sintering:* The preform is packed with ceramic powder and then heated to high temperatures (often below the melting point) to promote bonding and densification (increase in density) of the matrix.
- *Liquid infiltration:* Techniques like liquid silicon infiltration (LSI) involve infiltration with molten ceramic material, creating a strong interfacial bond with the fibers.

9.2.4 Consolidation and Finishing

- After infiltration, the CMC may undergo additional processes like hot pressing (applying high pressure and temperature) to further enhance its properties.
- Finally, the shaped CMC component undergoes machining and finishing steps to achieve the desired final dimensions and surface quality.

CMC processing can be complex and expensive due to the high temperatures involved and the need for precise control over fiber preform architecture. However, by meticulously crafting these processing steps, CMCs can be produced that bridge the gap between the exceptional strength of ceramics and the lightweight advantages of composites for various strategic applications [15].

9.2.5 Advanced CMC Manufacturing Processes

Challenges in traditional CMC processing techniques can be overcome with modern processing techniques such as additive manufacturing (AM). AM processes are utilized for fabricating various range of materials, including polymers, metals, ceramics, and composites. In comparison to polymers and metals, 3D printing to ceramics and CMCs is still in initial stages. Figure 9.3 shows the technological roadmaps of AM. AM techniques are stereolithography (SLA), selective laser sintering (SLS), laminated object manufacturing (LOM), inkjet printing (IJP), binder jetting (BJ), laser-engineered net shaping (LENS), direct ink writing (DIW), and hybrid additive manufacturing (H-AM). All the above-mentioned AM techniques have been explored in detail and reviewed in past years, which has been summarized in Table 9.1.

AM is three-dimensional printing (3DP), in which objects are constructed layer by layer from a digital model or CAD file. This technology comprises many techniques and substances, each presenting distinct benefits and practical uses. Typically, the 3DP procedure involves dividing the digital model into slender horizontal strata. The printer then applies material in a layer-by-layer fashion in accordance with the segmented data. The fusion of the layers may occur via a multitude of mechanisms, including melting, curing, or bonding,

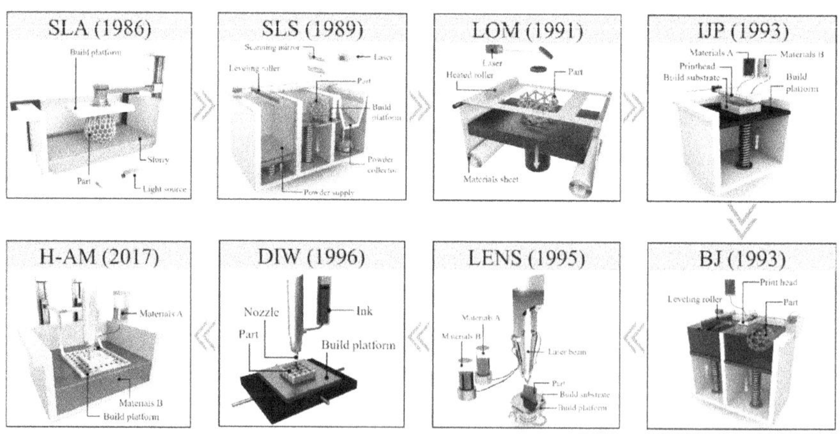

Figure 9.3 Roadmap of ceramic AM process. (Reprinted from Sun et al. [2023]. Copyright 2024, with permission from Elsevier.)

Table 9.1 Representative technology-specific reviews on ceramic AM

AM	Title of the review	Journal	Authors	Year
SLA	A comprehensive review or the photopolymerization of ceramic resins used in stereolithography	*Additive Manufacturing*	Zakeri et al.	2020
SLS	High-performance ceramic parts with complex shape prepared by selective laser sintering: A review	*Advances in Applied Ceramics*	Chen et al.	2018
LOM	laminated manufacturing of ceramic-based materials	*Advanced Engineering Materials*	Dermeik et al.	2020
BJ	Binder jetting of ceramics: Powders, binders, printing parameters. equipment. and post-treatment	*Ceramics International*	Lv et al.	2019
IJP	Additive manufacture of ceramics components by inkjet printing	*Engineering*	Derby et al.	2015
LENS	A review on laser deposition-additive manufacturing or ceramics and ceramic reinforced metal matrix composites	*Ceramics International*	Hu er al.	2018
DIW	Ceramic robocasting: Recent achievements, potential, and future developments	*Advanced Materials*	peng er al.	2018
H-AM	A review or digital manufacturing-based hybrid additive manufacturing processes	*The International Journal of Advanced Manufacturing Technology*	Chong et al.	2018

Source: Reprinted from Sun et al. (2023). Copyright 2024, with permission from Elsevier.

contingent upon the particular 3DP technique employed. Among the methods that are categorized as 3DP are the following: BJ, powder bed fusion (PBF), directed energy deposition (DED), sheet lamination, LOM, and so on. In addition to design flexibility, rapid prototyping, and the ability to produce complex geometries and customized parts on demand, 3DP provides many additional benefits. Application domains for this technology encompass aerospace, automotive, healthcare, and consumer goods, among others.

Despite all the methods mentioned above, SLA is inferior and utilizes a photo-curable liquid ceramic resin or paste to create a green part using a CAD model. A recoater attachment is typically required to apply a new slurry layer after curing subsequent layers [1, 16, 17]. The SLA technique is widely used and well regarded for its ability to produce precise CMC components due to its exceptional printing precision, surface quality, and relatively quick build speed [18]. Nevertheless, the fabrication of ceramic parts and reinforcing materials with higher refractive indices and having adequate light absorbing ability, such as carbides with reinforcements, is complex due to their limited cure depth when exposed to UV radiation. Moreover, an excessive quantity of organic material can cause a substantial reduction in solid loading, which in turn can result in distortion and the production of cracks during the debinding process. Hence, producing thick-walled components without encountering excessive porosity or breaking is challenging [19].

SLS is an AM process that utilizes a high-power laser to selectively fuse powders in order to fabricate components [20–23]. Direct metal laser sintering or selective laser melting (DMLS/SLM) is an AM technology in which the ceramic item is produced by directly melting it with a laser, eliminating the need for binder removal and sintering. Nevertheless, the mechanical characteristics of SLM-manufactured components cannot be compared to those achieved using indirect AM methods due to the susceptibility of ceramics to cracking. This is mostly due to ceramics having high melting temperatures, low thermal shock resistance, and low thermal conductivity [24]. The SLS technology is well regarded for its significant capability in producing diverse CMC components. Nevertheless, the parts produced by SLS may have significant porosity, resulting in low densities and inadequate mechanical characteristics. Additionally, SLS has less part resolution in comparison to SLA. Moreover, the SLS method is efficient due to its ability to print large layers, making it a reliable AM technology for creating porous composite components. This technique is particularly useful in producing bone tissue scaffolds and surface catalysts [25–27]. LOM is a method that involves gluing many layers of materials together to create 3D objects. This process includes removing unwanted elements before bonding occurs [28]. The LOM manufacturing technique prevents deformation by minimizing thermal stresses. In addition, LOM can produce monolithic as well as continuous fiber-reinforced ceramic composite components at a larger scale, which are difficult to manufacture using conventional AM processes [15]. Nevertheless, LOM is rarely extensively utilized because of its primary disadvantages, including material wastage, delamination, limited complexity in geometries, and anisotropic characteristics, in the planar axes [29]. BJ, also known as three-dimensional printing (3DP), is an AM process where liquid binder is carefully placed into the bed which is full of powder to form solid objects. Once the curing, drying, debinding, and sintering are completed, the finished pieces are obtained [30]. BJ has similar benefits to SLS, including rapid printing speed, the absence of a requirement for support structures, the ability to create porous parts easily, a diverse selection of materials, and a bigger printing volume [31]. However, the product produced by BJ exhibits porosity and generally comparatively inferior geometric precision as a result of the low density at which the powder is packed. Therefore, further processes after densification are required [27, 32]. Material jetting is an alternative term for IJP. During the procedure, a ceramic ink is evenly spread out to form ceramic pieces onto a build platform. Like other AM technologies such as BJ, DIW, and SLS, this technology is compatible with a wide range of materials [30, 32–39]. IJP is a potent microfabrication method because of its appreciable less surface roughness and great precision [28]. Moreover, it enables the production of functionally graded materials by accurately controlling the deposition of materials at the voxel level [40]. Although there are many benefits, the study of IJP of CMCs has not been done much due to its inconsistent and less reliable printing due to severe nozzle blockage [41]. LENS or direct energy deposition (DED), has been utilized to fabricate ceramic components by a process of layer-by-layer deposition. This technique involves the entire melting of ceramic powders, without using any binders, using a high-power laser beam. Like the direct SLM method, they possess a notable benefit over other AM techniques in that they are a one-step procedure that does not require heat treatment like debinding and sintering. The LENS method has significant benefits in terms of its high-resolution printing efficacy and the ability to reprint/recoat for repairing applications. Additionally, it allows for composition grading by employing several feeds; hence it can produce CMCs [42]. Nevertheless, the LENS method is hindered by its poor dimensional tolerances, high porosity, and restriction in terms of complex geometry, which ultimately restricts its commercialization. DIW, often referred to as robocasting [36], involves the direct extrusion of a highly viscous ceramic ink/paste or a high ceramic-loaded filament via a tiny opening in a layer-by-layer deposition process [43–45]. The DIW technique is more cost-effective and capable of constructing continuous fiber-reinforced CMCs due to its relatively simple setup and the absence of a costly heat source like a laser. DIW has an advantage in terms of entirely customized materials needs, and it now provides the widest range of printed materials. Furthermore, the addition of a strengthening component, such as chopped fibers, may be managed by adjusting the extrusion settings to achieve the desired orientation [73]. Nevertheless, the components' quality and resolution are much inferior to those generated using lithography-based method. The use of fillers creates the problem of jamming of nozzle, which in turn restricts the amount of reinforcement that can be included in the resulting composites [46–50]. H-AM refers to the combination of two or more 3D printing techniques. This approach allows for the creation of ceramic composite components with complex features that may not be feasible using individual processes alone. The integration of different methods opens up new possibilities for fabricating combination of materials and multifunctional ceramic parts. This has been shown to be economically advantageous compared to stand-alone processes. Peng et al. [51] used SLA and DIW to develop a combination AM approach for producing devices fulfilling several functions at the same time, in a one-go printing operation. Three-dimensional CERAM, a French business, has initiated efforts in Hybrid-AM methods by creating a hybrid 3D printer capable of producing low-temperature cofired ceramics made of ceramic and metal materials. Their H-AM technology utilizes a preexisting SLA 3D printer in conjunction with a hybrid extrusion tool. This hybrid 3D printer is fitted with an adding mechanism that allows for the incorporation of additional materials into the ceramic substrate during the printing process. Every AM process has demonstrated the potential to create CMCs. Nevertheless, there is no universally superior ceramic AM technology. Instead, the choice of the most appropriate technique depends on the specific ceramic composite being created. Hence, it is recommended that the selection of the AM technique for CMCs and their intended applications should be thoroughly evaluated, taking into account the notable benefits, constraints, and factors such as build volume and resolution. Figure 9.4 illustrates the conflicting relationship between construction volume and resolution. To achieve greater resolution in fabricating a part, it is necessary to decrease the build volume to a smaller size [21]. SLM and LENS methods do not require lengthy heat treatment; however, the significant temperature differences make it difficult to create components with high density. Additional 3D printing techniques, such as DIW, IJP, and BJ, offer the widest range of options for selecting printable materials. Hence, while considering specific ceramic materials and their intended uses, it is crucial to thoroughly examine all available AM processes in order to identify the most suitable alternatives [52].

This chapter also comprehensively reviewed the various AM processes for fabricating fiber-reinforced ceramic matrix composites (FRCMCs). The schematic diagram of various AM technologies for FRCMCs shown in Figure 9.5 includes SLA, digital light processing (DLP), SLS, BJ, fused deposition modeling (FDM), extrusion freeform fabrication (EFF), DIW, and three-dimensional printing (3DP). The SLA is a 3D printing technology that cures layers of photopolymer resin using a UV laser. SLA can fabricate components with intricate geometries and high resolution in FRCMCs. By incorporating fiber reinforcements into

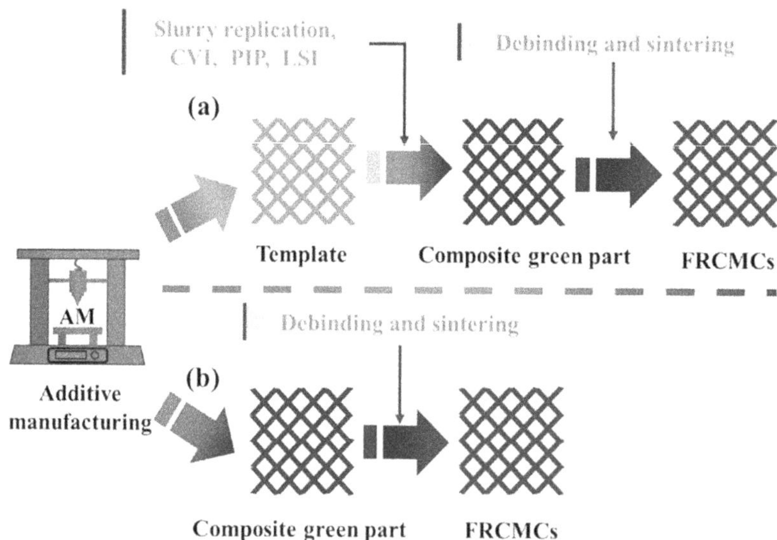

Figure 9.4 Indirect (a) and direct (b) AM processes [53]. (Reprinted from Wang et al. [2022]. Copyright 2024, with permission from Elsevier.)

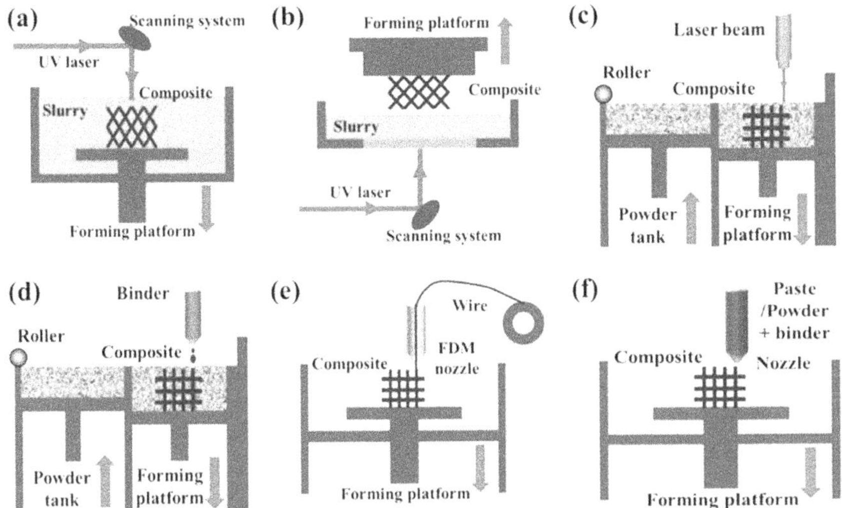

Figure 9.5 Schematic drawing of various AM technologies for FRCMCs: (a) SLA, (b) DLP, (c) SLS, (d) BJ, (e) FDM, and (f) EFF, DIW, 3DP. (Reprinted from Wang et al. [2022]. Copyright 2024, with permission from Elsevier.)

the resin, mechanical properties such as rigidity and strength can be improved. Comparable to SLA, DLP also employs UV-cured photopolymer resin [15].

In contrast, DLP cures entire strata simultaneously using a digital light projector instead of a laser. As a consequence, printing durations are reduced compared to SLA. Additionally, DLP can print FRCMC components with added fiber reinforcements. Using a high-powered laser, SLS entails layer-by-layer fusion of powdered materials, such as metals or polymers. FRCMCs are formed by sintering thermoplastic granules infused with fibers to produce strong and durable components. SLS provides design versatility and can fabricate intricate geometries devoid of support structures. BJ is a powder bed fusion method involving the selective deposition of a liquid binding agent onto a powder material layer. An application of this method to FRCMCs consists of the deposition of a binder onto a substrate of composite powder that incorporates fibers. The part is cured after printing to consolidate the binder and fuse the fibers, producing a composite part. FDM is a technique that produces items by constructing them from thermoplastic filaments layer by layer. The use of this approach for 3D printing is quite common. Continuous or fragmented fibers can be inserted into the thermoplastic matrix material in the context of FRCMCs while the printing process is being carried out. Through the use of FDM, it is possible to manufacture FRCMC components with intermediate mechanical qualities at an inexpensive price. EFF is an AM technique that creates three-dimensional objects by extruding materials to form layers. EFF is especially well-suited for the fabrication of large-scale components featuring intricate geometries. A material, typically a paste or semi-solid substance, is extruded onto a construction platform via a nozzle in EFF [41]. Computer-aided design (CAD) software regulates extrusion by directing the nozzle's motion to deposit the material in exact patterns. The layer endures a curing process with every successive deposition and solidifies, forming bonds with the preceding layer. Utilizing this layer-by-layer methodology permits the creation of complex geometries and internal frameworks. EFF is adaptable and can accommodate composites, metals, polymers, and ceramics, among other substances. EFF can be implemented in FRCMCs through the incorporation of fibers into the extruded material. By providing reinforcement, the fibers improve the mechanical properties of the composite ultimate product. One benefit of EFF is its capacity to manufacture components with a high material utilization efficiency, as waste is minimized by extruding only the required material. Moreover, in comparison to certain alternative AM methods, EFF can provide fabrication durations that are considerably quicker, rendering it well-suited for the rapid prototyping and mass production of components. DIW facilitates the controlled and precise deposition of materials to construct intricate three-dimensional entities. This technique is ideally adapted for fabricating objects featuring complex geometries, such as composite materials like fiber-reinforced composite matrix composites [54]. In DIW, a substance, frequently in the form of an ink or paste, is extruded onto a substrate or construct platform via a fine nozzle. The extrusion process is regulated by computer-aided design (CAD) software, which establishes the nozzle's trajectory for the sequential deposition of material. The deposited substance implements a curing or solidification procedure to adhere to the preceding layer, which culminates in forming the intended object. Fibers are commonly integrated into the ink or paste material employed in DIW to reinforce FRCMCs. These fibers may be assembled in a matrix material or aligned in particular orientations to attain desired mechanical properties, including strength, rigidity, and toughness. DIW provides numerous benefits for the production of FRCMC components [45, 55–59]:

- *Precision:* DIW permits precise control over materials' deposition, enabling the high-resolution fabrication of complex shapes and internal structures.
- *Personalization:* The adaptability of DIW permits the production of customized components that are designed to meet the needs of particular applications or specifications.
- *Diversity of materials:* DIW is compatible with an extensive array of materials, such as composites, metals, polymers, and ceramics, which renders it well-suited for a wide range of applications.
- *Orientation control:* DIW permits the distribution and orientation of fibers within the composite matrix to be regulated, thereby facilitating the enhancement of mechanical properties.
- *Layer-by-layer construction:* Similar to alternative AM methodologies, DIW constructs objects one at a time, thereby enabling the creation of parts with intricate internal structures without requiring supplementary tooling and granting design flexibility.

9.2.6 Summary and future scope

Various AM techniques—SLS, SLA, BJ, LENS, IJP, LOM, DIW, and H-AM—facilitate cutting-edge methods for creating CMCs. These techniques provide advantages such as the ability to produce complex shapes, customized production, cost-savings for small production runs, tailoring compositions and properties, and creating bioceramic composites with a customized reinforcement.

AM is revolutionizing the manufacturing industry and has garnered considerable attention from different strategic sectors such as biomedical, aerospace and space, new and sustainable energy, and automotive industries. Nevertheless, the utilization of end-used parts is currently restricted due to its early stage of development. Ceramic-based composites, such as customized zirconia/alumina composites, are commonly used in dental restoration. These composites are gaining popularity for replacing lost teeth due to their attractive appearance and ability to withstand damage. Another notable instance is the utilization of harsh environments. Effective production of components made of reinforced SiC include intricate channels for extreme-pressure turbine nozzles. CMCs created using AM were designed to address the challenges associated with traditional production methods. However, there is a need to investigate further and exploit their potential for a wider range of industrial applications. The adoption of AM in computerized numerical control machines is still early, and several challenges and concerns impede its widespread use in the industry. The mechanical durability of the composite is significantly constrained by various processing defects that may appear during printing of component or binder removal and sintering process. These defects include anisotropically dispersed porosity, microcracks, inadequate fusion or bonding, and nonuniform distribution of reinforcement. It is crucial to have exact control over these faults in order to create CMCs in a consistent and replicable way. Furthermore, the integration of continuous fiber reinforcement with AM poses a significant obstacle for researchers. There is an urgent need to create a novel AM technology that allows for the use of fiber-reinforced CMCs either on their own or in conjunction with other existing technologies. This is because the mechanical characteristics of fiber-reinforced CMCs are much superior to those of particle-reinforced CMCs. Furthermore, the capability to produce sizable CMC components with the required characteristics has not yet been demonstrated. Conventional AM techniques' limitations primarily lie in terms of build platform dimensions and dimensional accuracy with respect to surface roughness. BJ and DIW provide promising advantages for the production of large-sized components, while there is a requirement for substantial enhancement in the relative density of the manufactured parts. Furthermore, several applications for CMCs need exceptional dimensional precision. Nevertheless, the printing resolution of current AM processes restricts both the dimensional accuracy of parts and possible surface quality. CMC components fabricated by AM frequently exhibit substandard dimensional tolerance due to subsequent post-processing procedures, including drying, debinding, and sintering. As an illustration, when using the SLA method to produce as-sintered CMC pieces, significant shrinkage occurs, resulting in unpredictable dimensional accuracy. Furthermore, the presence of reinforcements during the SLA process might lead to light scattering, potentially impacting both printing accuracy and surface properties. Ultimately, bio-ceramic composites lack the natural ability of ceramic materials to withstand harsh conditions, thereby necessitating the investigation of both the metal-reinforced ceramics and ceramic-reinforced ceramic composites. With the hope that these issues may be resolved soon, we can expect the widespread use of AM processing for CMCs in the industrial sector.

REFERENCES

1. J. Sun, D. Ye, J. Zou, X. Chen, Y. Wang, J. Yuan, H. Liang, H. Qu, J. Binner, J. Bai, A review on additive manufacturing of ceramic matrix composites, J. Mater. Sci. Technol. 138 (2023) 1–16. https://doi.org/10.1016/j.jmst.2022.06.039.
2. O.L. Ighodaro, O.I. Okoli, Fracture toughness enhancement for alumina systems: A review, Int. J. Appl. Ceram. Technol. 5 (2008) 313–323. https://doi.org/10.1111/j.1744-7402.2008.02224.x.
3. E. Trunova, R. Herzog, T. Wakui, R.W. Steinbrech, E. Wessel, L. Singheiser, Micromechanisms affecting macroscopic deformation of plasma sprayed TBCs, in: 28th International Conference on Advanced Ceramics and Composites B: Ceramic Engineering and Science Proceedings, Wiley Online Library, 2008: pp. 411–416. https://doi.org/10.1002/9780470291191.ch62.
4. E. Rocha-Rangel, D. Hernández-Silva, E. Terrés-Rojas, E. Martínez-Franco, Alumina-based composites strengthened with titanium and titanium carbide dispersions, Epa. J. Silic. Based Compos. Mater. 62 (2010) 75–78. https://doi.org/10.14382/epitoanyag-jsbcm.2010.15.
5. G. Bansal, R.K. Gautam, J.P. Misra, A. Mishra, Physiomechanical, flowability, and antibacterial characterization of silver-doped eggshell-derived hydroxyapatite for biomedical applications, J. Mater. Eng. Perform. (2023). https://doi.org/10.1007/s11665-023-08696-6.
6. D. Sarkar, S. Adak, N.K. Mitra, Preparation and characterization of an Al_2O_3–ZrO_2 nanocomposite, Part I: Powder synthesis and transformation behavior during fracture, Compos. Part A Appl. Sci. Manuf. 38 (2007) 124–131. https://doi.org/10.1016/j.compositesa.2006.01.005.
7. A. Beitollahi, H. Hosseini-Bay, H. Sarpoolaki, Synthesis and characterization of Al_2O_3–ZrO_2 nanocomposite powder by sucrose process, J. Mater. Sci. Mater. Electron. 21 (2010) 130–136. https://doi.org/10.1007/s10854-009-9880-9.
8. P.K. Rao, P. Jana, M.I. Ahmad, P.K. Roy, Synthesis and characterization of zirconia toughened alumina ceramics prepared by co-precipitation method, Ceram. Int. 45 (2019) 16054–16061. https://doi.org/10.1016/j.ceramint.2019.05.121.
9. G. West, Modern ceramic engineering: Properties, processing and use in design, Compos. Manuf. 5 (1994) 58–59. https://doi.org/10.1016/0956-7143(94)90021-3.
10. G. Bansal, R.K. Gautam, J.P. Misra, A. Mishra, Tribological behavior of silver-doped eggshell-derived hydroxyapatite reinforcement in PMMA-based composite. (2024). https://doi.org/10.1177/14644207241240623.
11. G. Bansal, R.K. Gautam, J.P. Misra, A. Mishra, Coating methods for hydroxyapatite: A bioceramic material. (2023) 279–302. https://doi.org/10.1007/978-981-99-3549-9_13.

12. G. Bansal, R.K. Gautam, J.P. Misra, A. Mishra, Synthesis and characterization of poly(methyl methacrylate)/silver-doped hydroxyapatite dip coating on Ti6Al4V, Colloids Surfaces A Physicochem. Eng. Asp. (2024) 133662. https://doi.org/10.1016/j.colsurfa.2024.133662.

13. N. Singh, R. Mazumder, P. Gupta, D. Kumar, Ceramic matrix composites: Processing techniques and recent advancements, J. Mater. Environ. Sci. 8 (2017) 1654–1660.

14. M. Vlasova, B.S. Coeto, M. Kakazey, P.A.M. Aguilar, I. Rosales, A.E. Martinez, V. Stetsenko, A. Bykov, Laser synthesis of Al_2TiO_5 ceramics from Al_2O_3–TiO_2 powder mixtures, J. Ceram. Sci. Technol. 3 (2012) 61–68. https://doi.org/10.4416/JCST2012-00005.

15. M. Chen, H. Qiu, W. Xie, B. Zhang, S. Liu, W. Luo, X. Ma, Research progress of continuous fiber reinforced ceramic matrix composite in hot section components of aero engine, IOP Conf. Ser. Mater. Sci. Eng. 678 (2019) 012043. https://doi.org/10.1088/1757-899X/678/1/012043.

16. X. Zhang, K. Zhang, B. Zhang, Y. Li, R. He, Quasi-static and dynamic mechanical properties of additively manufactured Al_2O_3 ceramic lattice structures: Effects of structural configuration, Virtual Phys. Prototyp. 17 (2022) 528–542. https://doi.org/10.1080/17452759.2022.2048340.

17. J. Li, X. An, J. Liang, Y. Zhou, X. Sun, Recent advances in the stereolithographic three-dimensional printing of ceramic cores: Challenges and prospects, J. Mater. Sci. Technol. 117 (2022) 79–98. https://doi.org/10.1016/j.jmst.2021.10.041.

18. O. Gavalda Diaz, G. Garcia Luna, Z. Liao, D. Axinte, The new challenges of machining ceramic matrix composites (CMCs): Review of surface integrity, Int. J. Mach. Tools Manuf. 139 (2019) 24–36. https://doi.org/10.1016/j.ijmachtools.2019.01.003.

19. V. Morales-Flórez, A. Domínguez-Rodríguez, Mechanical properties of ceramics reinforced with allotropic forms of carbon, Prog. Mater. Sci. 128 (2022) 100966. https://doi.org/10.1016/j.pmatsci.2022.100966.

20. Y. Tang, X. Shen, Z. Liu, Y. Qiao, L. Yang, D. Lu, J. Zou, J. Xu, Corrosion behaviors of selective laser melted Inconel 718 alloy in NaOH solution, Jinshu Xuebao/Acta Metall. Sin. 58 (2022) 324–333. https://doi.org/10.11900/0412.1961.2021.00386.

21. K.S. Prakash, T. Nancharaih, V.V.S. Rao, Additive manufacturing techniques in manufacturing: An overview, Mater. Today Proc. 5 (2018) 3873–3882. https://doi.org/10.1016/j.matpr.2017.11.642.

22. P. Kocovic, 3D Printing and Its Impact on the Production of Fully Functional Components, IGI Global, 2017: pp. 1–25.

23. L.N. Zhang, O.A. Ojo, Corrosion behavior of wire arc additive manufactured Inconel 718 superalloy, J. Alloys Compd. 829 (2020) 154455. https://doi.org/10.1016/j.jallcom.2020.154455.

24. E. Feilden, C. Ferraro, Q. Zhang, E. García-Tuñón, E. D'Elia, F. Giuliani, L. Vandeperre, E. Saiz, 3D printing bioinspired ceramic composites, Sci. Rep. 7 (2017) 1–9. https://doi.org/10.1038/s41598-017-14236-9.

25. W.E. Frazier, Metal additive manufacturing: A review, J. Mater. Eng. Perform. 23 (2014) 1917–1928. https://doi.org/10.1007/S11665-014-0958-Z.

26. E.M. Sachs, J.S. Haggerty, M.J. Cima, P.A. Williams, Three-dimensional printing techniques, Google Patents (n.d.) 10. https://patents.google.com/patent/US5204055A/en (accessed November 12, 2021).

27. P. Kunchala, K. Kappagantula, 3D printing high density ceramics using binder jetting with nanoparticle densifiers, Mater. Des. 155 (2018) 443–450. https://doi.org/10.1016/j.matdes.2018.06.009.

28. Z. Chen, Z. Li, J. Li, C. Liu, C. Lao, Y. Fu, C. Liu, Y. Li, P. Wang, Y. He, 3D printing of ceramics: A review, J. Eur. Ceram. Soc. 39 (2019) 661–687. https://doi.org/10.1016/j.jeurceramsoc.2018.11.013.

29. M. Schwentenwein, J. Homa, Additive manufacturing of dense alumina ceramics, Int. J. Appl. Ceram. Technol. 12 (2015) 1–7. https://doi.org/10.1111/IJAC.12319.

30. Y. Lakhdar, C. Tuck, J. Binner, A. Terry, R. Goodridge, Additive manufacturing of advanced ceramic materials, Prog. Mater. Sci. 116 (2021) 100736. https://doi.org/10.1016/J.PMATSCI.2020.100736.

31. C. Hong, D. Gu, D. Dai, A. Gasser, A. Weisheit, I. Kelbassa, M. Zhong, R. Poprawe, Laser metal deposition of TiC/Inconel 718 composites with tailored interfacial microstructures, Opt. Laser Technol. 54 (2013) 98–109. https://doi.org/10.1016/j.optlastec.2013.05.011.

32. X. Lv, F. Ye, L. Cheng, S. Fan, Y. Liu, Binder jetting of ceramics: Powders, binders, printing parameters, equipment, and post-treatment, Ceram. Int. 45 (2019) 12609–12624.

33. A.D. Lantada, A. De Blas Romero, M. Schwentenwein, C. Jellinek, J. Homa, Lithography-based ceramic manufacture (LCM) of auxetic structures: Present capabilities and challenges, Smart Mater. Struct. 25 (2016) 054015. https://doi.org/10.1088/0964-1726/25/5/054015.

34. M. Lejeune, T. Chartier, C. Dossou-Yovo, R. Noguera, Ink-jet printing of ceramic micro-pillar arrays, J. Eur. Ceram. Soc. 29 (2009) 905–911. https://doi.org/10.1016/J.JEURCERAMSOC.2008.07.040.

35. B.A. Turtle, J.E. Smay, J. Cesarano, J.A. Voigt, T.W. Scofield, W.R. Olson, J.A. Lewis, Robocast $Pb(Zr_{0.95}Ti_{0.05})O_3$ ceramic monoliths and composites, J. Am. Ceram. Soc. 84 (2001) 872–874. https://doi.org/10.1111/J.1151-2916.2001.TB00756.X.

36. S. Tiwari, P.K. Singh, S. Kumar, Variation of density in additive manufacturing of ceramic parts: A review, Mater. Today Proc. 47 (2021) 2742–2745. https://doi.org/10.1016/j.matpr.2021.03.070.

37. M. Krinitcyn, Z. Fu, J. Harris, K. Kostikov, G.A. Pribytkov, P. Greil, N. Travitzky, Laminated object manufacturing of in-situ synthesized MAX-phase composites, Ceram. Int. 43 (2017) 9241–9245. https://doi.org/10.1016/J.CERAMINT.2017.04.079.

38. M. Allahverdi, S.C. Danforth, M. Jafari, A. Safari, Processing of advanced electroceramic components by fused deposition technique, J. Eur. Ceram. Soc. 21 (2001) 1485–1490. https://doi.org/10.1016/S0955-2219(01)00047-4.

39. S.B. Balani, S.H. Ghaffar, M. Chougan, E. Pei, E. Şahin, Processes and materials used for direct writing technologies: A review, Results Eng. 11 (2021) 100257. https://doi.org/10.1016/J.RINENG.2021.100257.

40. A. Zocca, P. Colombo, C.M. Gomes, J. Günster, Additive manufacturing of ceramics: Issues, potentialities, and opportunities, J. Am. Ceram. Soc. 98 (2015) 1983–2001. https://doi.org/10.1111/jace.13700.

41. B. Khatri, K. Lappe, M. Habedank, T. Mueller, C. Megnin, T. Hanemann, Fused deposition modeling of ABS-barium titanate composites: A simple route towards tailored dielectric devices, Polymers 10 (2018) 666. https://doi.org/10.3390/POLYM10060666.

42. S. Siengchin, A review on lightweight materials for defence applications: Present and future developments, Def. Technol. 24 (2023) 1–17. https://doi.org/10.1016/j.dt.2023.02.025.

43. J.A. Lewis, J.E. Smay, J. Stuecker, J. Cesarano, Direct ink writing of three-dimensional ceramic structures, J. Am. Ceram. Soc. 89 (2006) 3599–3609. https://doi.org/10.1111/J.1551-2916.2006.01382.X.

44. U. Golcha, A.S. Praveen, D.L. Belgin Paul, Direct ink writing of ceramics for bio medical applications: A review, IOP Conf. Ser. Mater. Sci. Eng. 912 (2020) 032041. https://doi.org/10.1088/1757-899X/912/3/032041.

45. N.W.S. Pinargote, A. Smirnov, N. Peretyagin, A. Seleznev, P. Peretyagin, Direct ink writing technology (3D printing) of graphene-based ceramic nanocomposites: A review, Nanomaterials. 10 (2020) 1–48. https://doi.org/10.3390/nano10071300.

46. X. Xu, J. Zhang, P. Jiang, D. Liu, X. Jia, X. Wang, F. Zhou, Direct ink writing of aluminum-phosphate-bonded Al_2O_3 ceramic with ultra-low dimensional shrinkage, Ceram. Int. 48 (2022) 864–871. https://doi.org/10.1016/j.ceramint.2021.09.168.

47. X. Xu, M. Zhang, P. Jiang, D. Liu, Y. Wang, X. Xu, Z. Ji, X. Jia, H. Wang, X. Wang, Direct ink writing of Pd-decorated Al_2O_3 ceramic based catalytic reduction continuous flow reactor, Ceram. Int. 48 (2022) 10843–10851. https://doi.org/10.1016/j.ceramint.2021.12.301.

48. Z. Li, J. Li, H. Luo, X. Yuan, X. Wang, H. Xiong, D. Zhang, Direct ink writing of 3D piezoelectric ceramics with complex unsupported structures, J. Eur. Ceram. Soc. 42 (2022) 3841–3847. https://doi.org/10.1016/j.jeurceramsoc.2022.03.038.

49. G. Franchin, P. Scanferla, L. Zeffiro, H. Elsayed, A. Baliello, G. Giacomello, M. Pasetto, P. Colombo, Direct ink writing of geopolymeric inks, J. Eur. Ceram. Soc. 37 (2017) 2481–2489. https://doi.org/10.1016/j.jeurceramsoc.2017.01.030.

50. A. M'Barki, L. Bocquet, A. Stevenson, Linking rheology and printability for dense and strong ceramics by direct ink writing., Sci. Rep. 7 (2017) 6017. https://doi.org/10.1038/s41598-017-06115-0.

51. X. Peng, X. Kuang, D.J. Roach, Y. Wang, C.M. Hamel, C. Lu, H.J. Qi, Integrating digital light processing with direct ink writing for hybrid 3D printing of functional structures and devices, Addit. Manuf. 40 (2021) 101911. https://doi.org/10.1016/j.addma.2021.101911.

52. G. Franchin, L. Wahl, P. Colombo, Direct ink writing of ceramic matrix composite structures, J. Am. Ceram. Soc. 100 (2017) 4397–4401. https://doi.org/10.1111/jace.15045.

53. W. Wang, L. Zhang, X. Dong, J. Wu, Q. Zhou, S. Li, C. Shen, W. Liu, G. Wang, R. He, Additive manufacturing of fiber reinforced ceramic matrix composites: Advances, challenges, and prospects, Ceram. Int. 48 (2022) 19542–19556. https://doi.org/10.1016/j.ceramint.2022.04.146.

54. L. Biasetto, G. Franchin, H. Elsayed, G. Boschetti, K. Huang, P. Colombo, Direct ink writing of cylindrical lattice structures: A proof of concept, Open Ceram. 7 (2021) 100139. https://doi.org/10.1016/j.oceram.2021.100139.

55. G. Wang, Y. Miao, H. Gong, M. Sheng, J. Jing, M. Liu, J. Lu, Z. Gong, K. Ma, Direct ink writing of reaction bonded silicon carbide ceramics with high thermal conductivity, Ceram. Int. 49 (2023) 10014–10022. https://doi.org/10.1016/j.ceramint.2022.11.179.

56. T. Lin, J. Duan, J. Wu, H. Shao, Direct ink writing of dense strontium ferrite parts by staggered and overlapped stacking method, Mater. Lett. 323 (2022) 132546. https://doi.org/10.1016/j.matlet.2022.132546.

57. Z. Chen, Z. Xu, F. Cui, J. Zhang, X. Sun, Y. Shang, R. Guo, N. Liu, S. Cai, C. Zheng, Direct ink writing of cordierite ceramics with low thermal expansion coefficient, J. Eur. Ceram. Soc. 42 (2022) 1685–1693. https://doi.org/10.1016/j.jeurceramsoc.2021.12.017.

58. M.S. Hassan, S. Zaman, A. Rodriguez, L. Molina, C.E. Dominguez, Y. Lin, Additive manufacturing of highly flexible impact wave propagation sensor, in: A. Wissa, M. Gutierrez Soto, R.W. Mailen (Eds.), Behavior and Mechanics of Multifunctional Materials XVII, SPIE, 2023, p. 14. https://doi.org/10.1117/12.2664163.

59. A. Motealleh, S. Eqtesadi, F.H. Perera, A.L. Ortiz, P. Miranda, A. Pajares, R. Wendelbo, Reinforcing 13–93 bioglass scaffolds fabricated by robocasting and pressureless spark plasma sintering with graphene oxide, J. Mech. Behav. Biomed. Mater. 97 (2019) 108–116. https://doi.org/10.1016/j.jmbbm.2019.05.016.

Chapter 10

Implementation of biomimicry for advanced impact-resistant composites: Advanced manufacturing techniques

Shreya Rai

10.1 INTRODUCTION TO BIOMIMICRY AND ITS IMPORTANCE IN COMPOSITES

In the course of thousands of years, nature has gradually evolved to the current state and in this process, it has found ways to exist and function most efficiently and sustainably. It has achieved this by making optimum use of the available resources and has also thrived in extreme conditions. Studying the remarkable achievement is essential, if not crucial, for incorporating its marvels into engineering designs. This study of drawing inspiration from structures, materials, phenomena, and processes occurring in nature, understanding their principle of working, and applying it in the fields of science, engineering, and medicine is known as biomimicry or bioinspiration. This bioinspiration is used in a wide range of applications like microfluids (inspired by veins and arteries in living organisms), soft robotics (robot structure inspired by starfish to pick up objects), flapping aircraft wings (inspired by birds), nano-optics, and others.

Strength and toughness are the two most desired properties of a structural material. However, in engineering materials, they are mostly found to be mutually exclusive. As the strength of a material increases, so does its brittle nature, and it is not easy to achieve properties of both high strength and high toughness in the same material. However, this is not the same for biological materials. They are known to show both the property simultaneously. Biological materials are composite materials of hard and strong biominerals and resilient and toughness-imparting biopolymers. This gives a massive benefit over conventional engineering materials. Hence, it is essential to study the biological materials and incorporate their features to create high-performance composite materials. Figure 10.1 shows that Young's modulus is low and toughness is high for biopolymers, while the stiffness is high but toughness is low for the mineral phase.

Such composites of hard and soft phases possessing remarkable impart resistance are abundant in nature, for example, the nacre structure in shells, the exoskeleton of beetles and crabs, the structure of bones, the structure of bamboo, and so on. Inspirations from various efficiency-performing natural structures

DOI: 10.1201/9781003564355-10

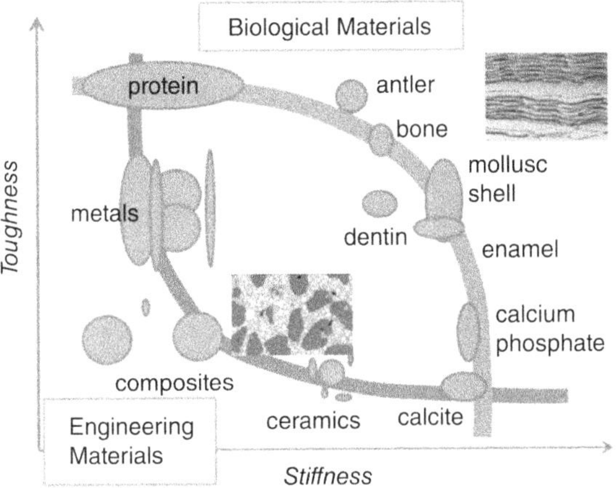

Figure 10.1 Toughness versus stiffness in biological materials and engineering materials [1].

have been incorporated into the structural aspect of engineering to gain superior impact resistance. For example, the tough protective covering of an armadillo is studied, and its structure is used in designing better performing body armour. This structure, when built in composites, can help us achieve composites with much better impact resistance. In this chapter, we study the design principles of nature and its implementation in composites. Next, we discuss the natural structures of superior impact resistance and its toughening mechanisms. Later, methods of fabrication of bioinspired composites are also discussed.

10.2 BIOMIMICRY DESIGN PRINCIPLES AND THEIR APPLICABILITY IN COMPOSITES

In order to design bioinspired materials, we need to understand the design principles followed by the nature in building biological materials and structures. These design principles are unique to nature and can be seen in majority of its creations.

10.2.1 Nature uses only the energy it needs and relies on freely available energy

Nature uses only the energy available to it. The resources used by the nature are the readily available resources, and does not require mining or any other processes to be extracted. All the processes of nature occur at ambient temperature and pressure and the energy required for these processes are freely available energy like the solar energy, wind energy, tidal energy, and so on. Energy requirements is a constraint in most of the design problems. To optimise the design output, the available form of energy is sought after for an efficient and sustainable design.

10.2.2 Nature recycles all materials

In nature, the by-product of one process becomes raw material for the other process. This process of recycling ensures optimal use of the resources and less wastage, which are some of the objectives of sustainable design. Use of biodegradable fibres, natural fibres, and biodegradable polymers in composite can take use closer to the sustainability goals.

10.2.3 Nature is resilient to disturbances

To be resilient is to have minimal or no effect in function on occurrence of external damaging factors. This quality can be seen in nature in the form of redundancy, decentralisation, self-renewal, and self-repair. Designing self-healing polymers or fibres in composite can help build advance composite technologies.

10.2.4 Nature tends to optimise rather than maximise

A well-engineered design is optimising the resources and design as per required outputs and not maximising. This property of optimising can be seen in nature. Customisation of composites structures by changing the design as per required characteristics can be one such way. This can be addressed by making use of designs like sandwich composites, or changing the fibre arrangement, material, number of layers for specific characteristic, and so on.

10.2.5 Nature provides mutual benefits

Various processes and entities of nature are interdependent or co-dependent on each other. Symbiosis is one such example where both the entities benefit from the co-dependence.

10.2.6 Nature runs on information

Nature uses the information of its environment to constantly evolve towards better functionality. Evolution is an exceptional process of collecting information from the surrounding and adapting to the environment, improving efficiency, and being more resilient and imperishable. The idea of collecting data from the past working and evolving existing technology to better, advanced technology can be incorporated in current engineering designs.

10.2.7 Nature uses chemistry and materials that are safe for living beings

The raw materials and chemical resources used for the processes in nature are non-toxic, biodegradable, and harmless to living entities. Adopting this principle helps towards sustainable goals.

10.2.8 Nature builds using abundant resources, incorporating rare resources only sparingly

Nature builds on materials readily available and abundant to it. It uses nitrogen, oxygen, and carbon in abundance. These abundant resources do not need any process to be mined or refined.

10.2.9 Nature uses shape to determine functionality

Nature determines the functionality of an entity by its shape rather than any added mechanisms of energy. Various such examples are seen in nature, like the bright blue colour of the butterfly wings is a result of interaction of light with the microscopic structure on its wing, the hydrophobicity of the lotus leaves arises due to structure and not any additional surface. The beetle's back is able to collect water during harsh weather conditions due to its structure. Learning these characteristics of natural materials, we can implement these in advanced composite materials to achieve higher surface friction, hydrophobicity, and so on just with the microstructures.

Implementing these principles will ensure an overall efficient design with optimised use of resources and superior functionality. Incorporating these design principles in advance composite designs can help us achieve more efficient, economic, and sustainable alternatives to the existing designs.

10.3 IMPLEMENTATION METHODS[2]

The approaches for biomimicry can be broadly classified into two types: the bottom-up approach and the top-down approach. In the top-down approach, designers start with a specific problem they want to solve and then look to nature for existing solutions. They analyse biological systems, organisms, or processes that exhibit behaviours or characteristics relevant to their problem. By studying these natural models, designers derive inspiration to develop innovative solutions. For example, a designer who wants to create a self-cleaning surface might study the lotus leaf, which repels water and prevents dirt from sticking. The designer can develop materials or coatings with similar self-cleaning capabilities by understanding the lotus leaf's microstructure and hydrophobic properties.

In the bottom-up approach, the initial focus lies on identifying an appropriate biological system of interest. Through in-depth examination of these biological systems, researchers uncover phenomena that may not have been previously apparent. Subsequently, efforts are made to integrate this understanding of biological systems into addressing a specific target need.

10.4 ADVANCED BIOLOGICAL DESIGN FEATURES

10.4.1 Layered (brick and mortar) [3, 4]

The layered structure found in nature are beneficial for improving the toughness of materials, especially the brittle ones. These structures contain multiple layers or interfaces. They are found in the nature in the form of layer of abalone shell, scales of fishes, and exoskeleton of beetles. Abalone shells consist of a unique brick and mortar-layered structure known as the nacre structure. The brick structure of nacre is primarily made up of minerals, namely aragonite ($CaCO_3$) bonded by protein-based thin layer (20–30 nm). Minerals are weak in toughness; however, this composite of nacre structure has exceptionally high toughness compared to either of its constituents, owing to its structure. The introduction of protein interface between the mineral platelets provides improved toughness, as they allow inelastic deformation and redistribution of stresses. The toughness mechanisms of the nacre structure include can also be understood better:

- The breaking of aragonite mineral bridges which exists between the mineral layers.
- The inelastic shearing resisted by the protein interface.
- The organic protein layer acting as viscoelastic glue, which involves breakage of sacrificial bonds in the biopolymer at molecular level.
- Tablet interlocking due to sliding.

Thus, the layered composites function to improve the toughness of the structure in many folds compared to the constituent materials (Figure 10.2).

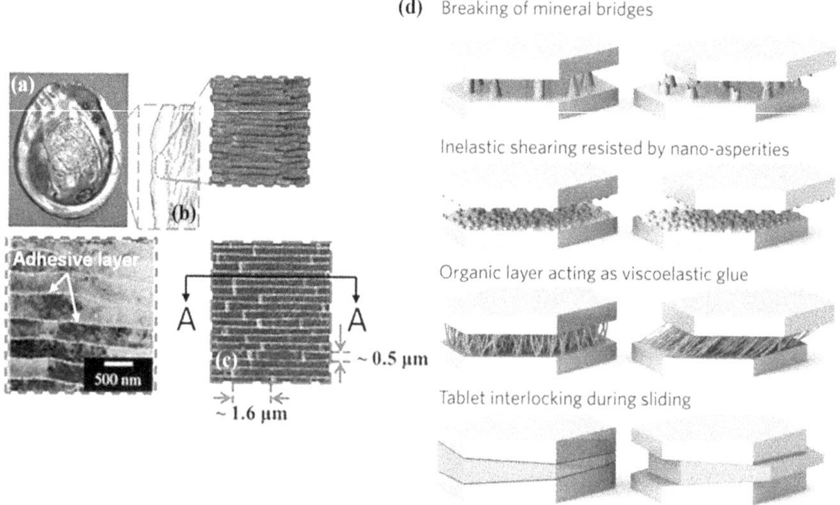

Figure 10.2 (a) Red abalone seashell. (b) The protective armour system. (c) The bricks and mortar microstructured. (d) Toughening mechanisms in nacre tablets [5, 6].

Introduction of layers and inelastic bonding between them can help improve the toughness of brittle materials to a great extent.

10.4.2 Helicoidal structure [7, 8]

Helicoidal structure, also known as Bouligand structure or twisted plywood structure or chiral nematic, or cholesteric, is a stacking sequence where the fibre layer is skewed at an angle called pitch angle with respect to the layer adjacent to it. This helicoidal structure was first observed by Bouligand in 1972. This Bouligand structure is now found in several organisms like arapaima (*Arapaima gigas*), sheep crab, peacock mantis shrimp (*Odontodactylus scyllarus*), claws of lobster, bone of mammals, the exoskeleton of beetles, dactyl club of mantis shrimp, armadillo osteoderm, and scales of carp fish and coelacanth fish. The dactyl club of mantis shrimp (Stomatopoda), which belongs to the species of marine crustaceans, has an exceptional damage tolerance as it hits its prey with multiple high-velocity strikes (over 20 m/s) without structural failure. These structures feature a twisted stair-like stacked lamellar structure. Each lamella consists of fibre having a diameter of 50–300 nm, which is composed of aligned chitin fibrils with a diameter of 3 nm wrapped in proteins.

Research conducted on these structures suggests that fibrous hierarchical arrangements have the ability to absorb significant deformation by spreading out applied loads and absorbing surplus energy. Moreover, they effectively arrest any local fractures, thereby augmenting their toughness, as shown in Figure 10.3. These structures demonstrate high toughness and resistance to impacts due to various toughening mechanisms, such as deflecting cracks, reorienting fibres, twisting, stretching, delamination, and deforming the fibrous matrix.

10.4.3 Bone-like structures [9]

The function of the bone is to act as a stiffening frame to the body and to impart toughness on possible occurrence of impact. This bone-like structure is present in various biological animals too, like the antlers of the deer or elk,

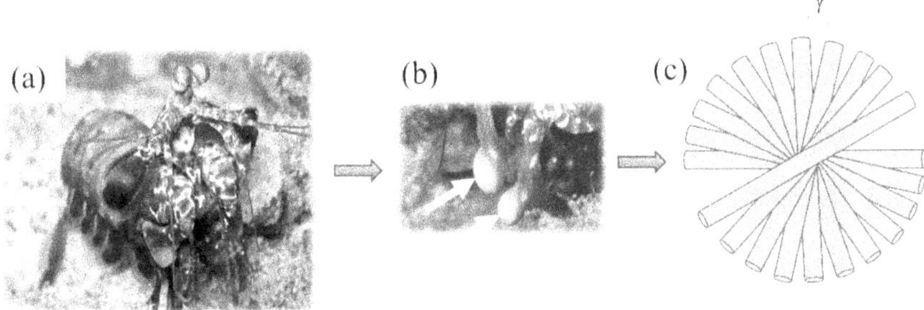

Figure 10.3 (a) Peacock mantis shrimp. (b) Dactyl club of mantis shrimp. (c) Schematic diagram of helicoidal structure of fibre arrangement in dactyl club of peacock mantis shrimp.

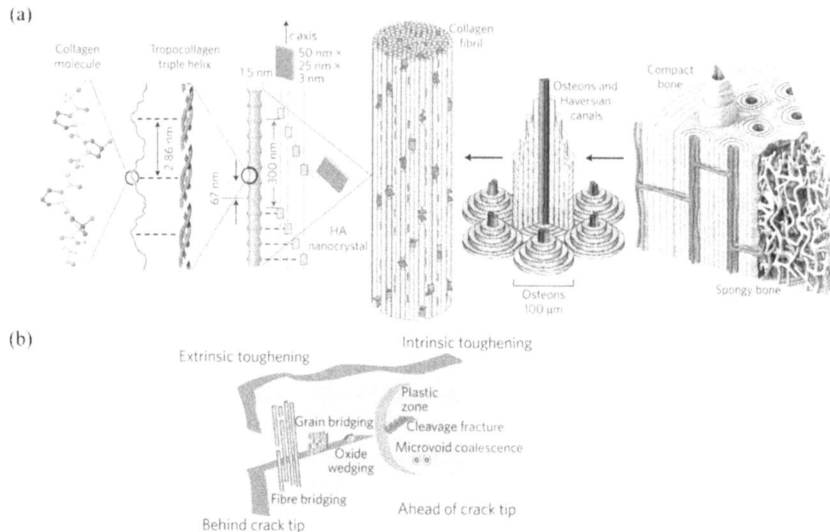

Figure 10.4 (a) Hierarchical structure of a bone. (b) Toughening mechanisms in a bone [6].

where they are subjected to high impacts of about 19 kJ. It is imperative to study about the structure of bone and their toughening mechanisms.

A bone structure is a composite of proteins (collagen fibres) and mineral nanoparticles (hydroxyapatite crystal). It is a multifunctional structure known to resist fracture at multiple scales. The bone structure shows hierarchy at different levels, it shows lamellar structure of fibres at nanoscale level and Haversian structures (osteons) at microscale level, as shown in Figure 10.4. The brittle hydroxyapatite mineral crystals made up of calcium and phosphate give bone its firmness and the protein collagen give the bone its toughness and strength through its ductility and fibrous nature.

The primary toughening mechanism is the breaking of the sacrificial bonds in the fibrils. This brings about energy dissipation and improves the toughness. Various other toughening mechanisms play an important role such as crack deflection, cracking bridging, viscoelastic flow, and constrained microcracking along the interfaces of osteons. The numerous microcrack formation parallel to the direction of the fracture accounts for toughening by uncracked ligament bridging. The crack deflection causes local crack arrest and increase in crack length, thereby increasing the overall toughness. Thus, the bone structure shows toughening mechanisms at various length scales.

10.4.4 Suture [10]

Suture structures are intricate arrangement of bones, plates, or scutes with a zigzag or wavy interfaces in between them. These structures are found as the outer covering of various organisms like the turtle-back, armadillo osteoderms, boxfish scales, and so on. The suture interface varies geometrically across different species. The function of these structures is to impart strength as well as flexibility to the outer covering. Another function of these structures is to withstand high-magnitude high-strain-rate impacts from the external environment. The constituents of the suture structure are similar to the bone structure with strength imparting biominerals as plates or scutes and flexible interlocking biopolymers or collagen fibres. Owing to this nature of the suture structures, they are able to absorb high energy resulting from impact. The sutures or the interphase of the bony plates are devoid of minerals, and hence provide extra flexibility to the structure.

10.4.5 Tubular [4, 11]

Tubular structures are a fascinating feature found in various materials, where long, aligned pores create a unique array within the material's bulk. These structures are commonly observed in natural materials that require resilience against impacts, such as hooves, teeth, and fish scales. Examples of materials featuring tubules include keratin-based horse hooves, ram horns, crab exoskeletons made of chitin, and human teeth composed of collagen and hydroxyapatite.

The significance of tubules goes beyond mere structural support; they play a crucial role in enhancing the toughness and energy absorption capabilities of these materials, as can be observed from Figure 10.5. For instance, in whale baleen—a filtering system in the mouths of baleen whales—the presence of tubules helps in resisting fractures and absorbing energy, particularly due to its low mineralisation level.

Mechanically, tubular structures improve the fracture toughness of materials by various means. They can impede the growth of cracks by removing stress concentration at the crack tip or by collapsing when subjected to compression forces. Moreover, they act as scattering centres, reducing the amplitude of stress waves generated by impacts, which is particularly vital in tissues like hooves and horns experiencing high-velocity loads.

When subjected to radial compression along the minor axis, tubules can deform and close, whereas loading along the major axis leads to distortion.

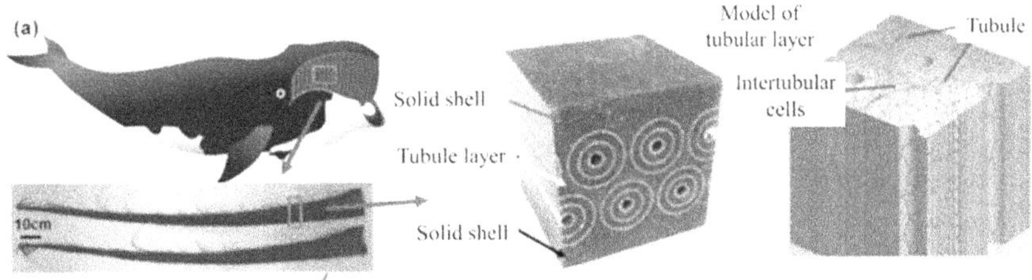

Figure 10.5 (a) Whale's baleen. (b) Model of tubular structure [12].

Additionally, during shear deformation, tubules may coalesce and form fibre bridges. Under compression or high-strain-rate impacts, tubules and lamellae within the material can buckle and crack. These deformations, including closing, coalescing, buckling, and cracking, play a crucial role in dissipating energy. This energy dissipation is particularly significant in scenarios such as the fighting behaviour of big-horn sheep, where high-strain-rate impacts occur.

10.5 TECHNIQUES TO MANUFACTURE ADVANCED BIOINSPIRED COMPOSITES

10.5.1 Hand lay-up and vacuum bagging (helical)

The process of hand lay-up and vacuum bagging can be used to fabricate bioinspired helicoidal FRP laminate composites (Figure 10.6). The process is mentioned as follows:

- Select continuous fibre and epoxy matrix for fabrication of FRP composite.
- Design the helicoidal structure by choosing number of layers and pitch angle.
- Cut the fibre into rectangular sheets at required angle.

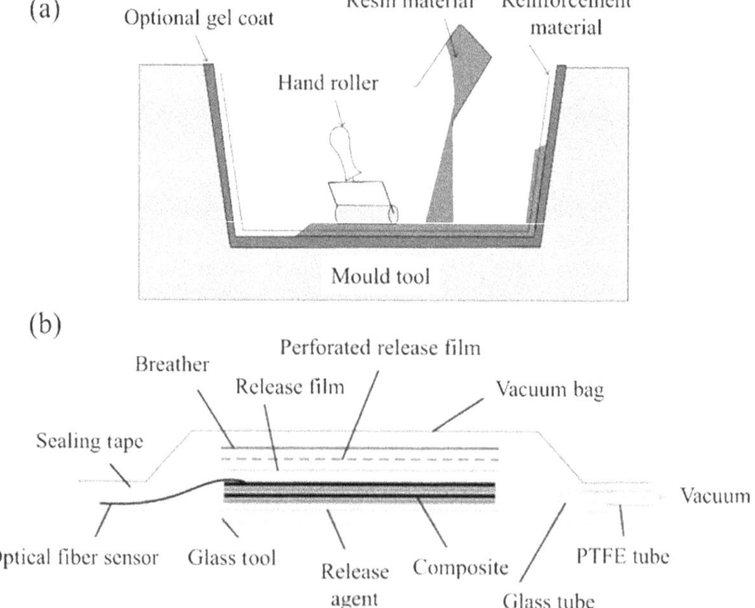

Figure 10.6 (a) Hand lay-up technique. (b) Schematic diagram of a vacuum bag technique [13, 14].

- Mix the epoxy and hardener in 10:1 ratio and lay the fibre sheets and spread this mixture uniformly. Remove excess epoxy by rolling.
- Cure the laminate in a vacuum bag at suitable pressure to eliminate any air voids.

10.5.2 Hand lay-up (tubular) [15]

Schematic diagram of bioinspired tubular structured composite is shown in Figure 10.7.

- Choose a suitable core material and external and internal tubes. One example is glass fibre as core material and aluminium as internal and external tubes. Design the number of layers and the direction of fibres as per requirements.
- Fabricate composite tubes using a roll-wrapping process. This involves wrapping the chosen fibre sheets around a mandrel.
- This process of manual laying up method can be improved by exerting mechanical pressure on the wrapped layers during fabrication.
- Overwrap Dacron fibre sheets around GFRP fibres to remove excess resin trapped during manufacturing, ensuring a smooth and clean finished surface.
- To avoid imperfections or delamination, apply pressure to each specimen using belts installed on waxed semi-polymer pipes in 10-cm segments.
- Cure the samples for 24 hours and then subject them to a post-cure process at room temperature for one week to ensure sufficient strength.

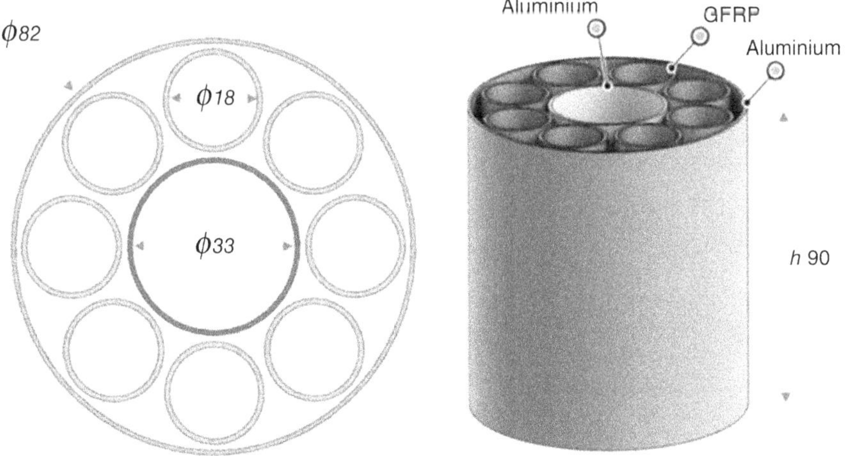

Figure 10.7 Schematic diagram of bioinspired tubular structured composite [15].

10.5.3 Powder Metallurgy

The process of power metallurgy can be used to fabricate composites with nacre structure. The process is described as follows:

- First, the constituent powders (e.g. Ti and Al) is milled to flake-like shapes using high-energy ball milling.
- Flakes of Ti, Al, and graphene nanosheets (GNSs) are then mixed by low-energy ball milling.
- Composite powders are then arranged in a graphite mould, forming a laminated structure.
- Composite powder are uniaxially compacted in the mould.
- Mould with compacted powder is sintered under a vacuum and a pressure.
- Sample is then cooled to room temperature before retrieval from the vacuum furnace.

The fabrication process of bioinspired nacre design using powder metallurgy is shown in Figure 10.8.

10.5.4 Additive manufacturing [17]

The process of additive manufacturing or 3D printing can be used to fabricate suture inspired composites as follows:

- Selection of two base materials, one hard and stiff and the other rubbery and elastic for the interface to mimic the constituent materials of suture structures.

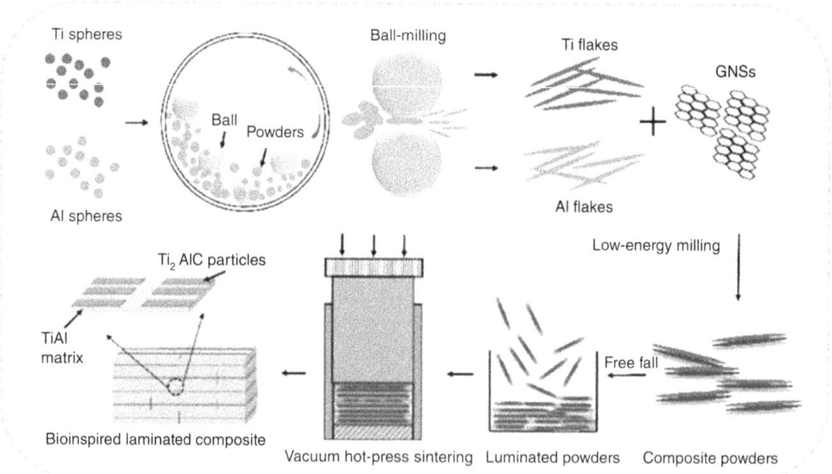

Figure 10.8 Typical block diagram fabrication process of bioinspired nacre design using powder metallurgy [16].

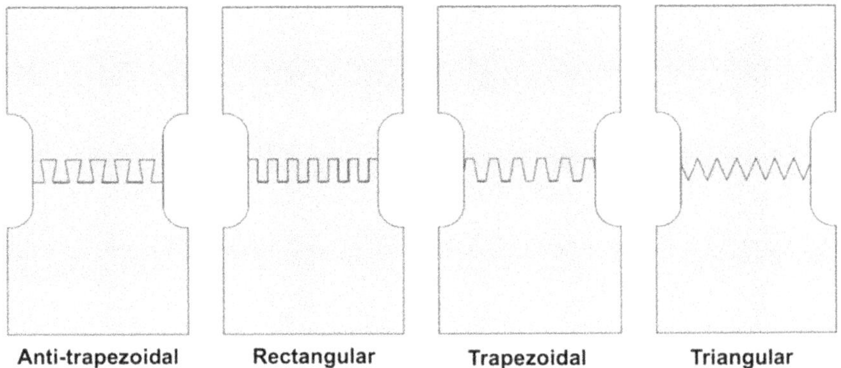

Figure 10.9 Schematic of 3D-printed bioinspired suture interface with different tailored waveforms (anti-trapezoidal, rectangular, trapezoidal, and triangular) [18].

- Designing of the composite in terms of number of layers and geometry of the interface.
- Model the design using 3D printing software and 3D print the design with multi-material 3D printer.

With the use of 3D printing, various complex geometry of suture interfaces can be printed and studied. Schematic of 3D-printed bioinspired suture interface with different tailored waveforms is shown in Figure 10.9. The technology of additive manufacturing enables to print various such complicated bioinspired structures like the bone-inspired Haversian structure, nacreous structures, structure of fish-scales, and so on.

10.5.5 Slip casting [5]

Th process of hot-press assisted slip casting (HASC) effectively combines hot pressing and slip casting and is used to manufacture nacre-inspired composites. The schematic diagram of slip casting process of fabrication of nacre-inspired ceramic composites is shown in Figure 10.10. The process takes place in the following steps:

- A suitable reinforcement fibre is selected with an aspect ratio similar to that of nacre—for example, alumina (Al_2O_3) flake-reinforced *epoxy matrix* nanolaminar composites.
- The reinforcement fibre/particle is mixed with a suitable amount of epoxy resin and hardener. Also, it is necessary to ensure the volume fraction is less than the highest practical content to maintain a manageable viscosity.
- Prepare a disposable porous plaster mould and position it at the bottom of a stainless steel die.
- The reinforcement particle and epoxy resin mixture is poured into the steel die, ensuring even distribution.

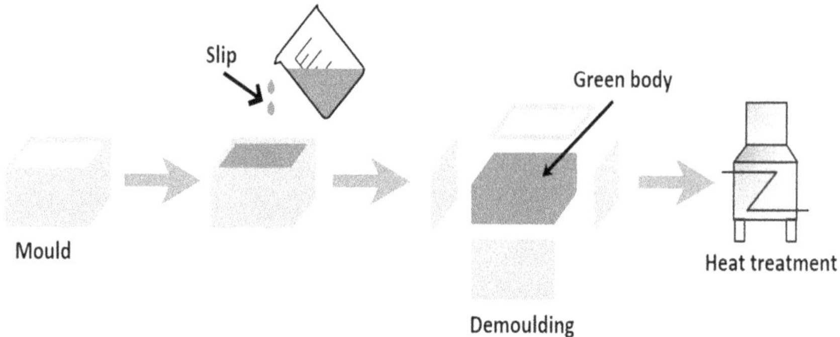

Figure 10.10 Schematic diagram of slip casting process of fabrication of nacre-inspired ceramic composites [19].

- Excess resin is pushed through the porous mould by applying an initial pressure to the die assembly.
- The die assembly is then placed into a custom-made laboratory hot press. The material is cured by maintaining temperature and pressure at the desired level.
- After curing, the epoxy matrix is cooled in ambient air at the end of the cycle.

10.6 CONCLUSION

Incorporating biomimicry principles into the design and manufacturing of advanced composites holds significant promise for enhancing impact resistance and toughness. By studying nature's ingenious solutions to structural challenges, engineers can unlock new possibilities for creating materials that are both efficient and sustainable. The diverse range of natural structures, from layered composites to tubular arrangements, offers valuable insights into toughening mechanisms that can be replicated in synthetic materials. Moreover, advanced manufacturing techniques such as additive manufacturing and powder metallurgy enable the precise fabrication of complex biomimetic designs. By embracing biomimicry, we can develop composites that not only meet the demands of modern engineering but also contribute to a more sustainable future.

REFERENCES

1. Gao W, Zhang Y, Ramanujan D, et al. The status, challenges, and future of additive manufacturing in engineering. *Computer-Aided Design.* 2015;69:65–89. doi: https://doi.org/10.1016/j.cad.2015.04.001
2. Aziz MS, El Sherif AY. Biomimicry as an approach for bio-inspired structure with the aid of computation. *Alexandria Engineering Journal.* 2016;55(1):707–714. doi: https://doi.org/10.1016/j.aej.2015.10.015

3. Ritchie RO. The conflicts between strength and toughness. *Nature Materials.* 2011;10(11):817–822. doi: 10.1038/nmat3115

4. Naleway SE, Porter MM, McKittrick J, Meyers MA. Structural design elements in biological materials: Application to bioinspiration. *Advanced Materials.* 2015;27(37):5455–5476. doi: https://doi.org/10.1002/adma.201502403

5. Le TV, Ghazlan A, Ngo T, Nguyen T, Remennikov A. A comprehensive review of selected biological armor systems: From structure–function to biomimetic techniques. *Composite Structures.* 2019;225:111172. doi:10.1016/j.compstruct.2019.111172

6. Wegst UGK, Bai H, Saiz E, Tomsia AP, Ritchie RO. Bioinspired structural materials. *Nature Materials.* 2015;14(1):23–36. doi: 10.1038/nmat4089

7. Raabe D, Sachs C, Romano P. The crustacean exoskeleton as an example of a structurally and mechanically graded biological nanocomposite material. *Acta Materialia.* 2005;53(15):4281–4292. doi: https://doi.org/10.1016/j.actamat.2005.05.027

8. Behera RP, Le Ferrand H. Impact-resistant materials inspired by the mantis shrimp's dactyl club. *Matter.* 2021;4(9):2831–2849. doi: https://doi.org/10.1016/j.matt.2021.07.012

9. Chen PY, Novitskaya E, Lopez MI, Sun CY, McKittrick J. Toward a better understanding of mineral microstructure in bony tissues. *Bioinspired, Biomimetic and Nanobiomaterials.* 2014;3(2):71–84. doi: 10.1680/bbn.13.00017

10. Shahar R, Kraus S, Monsonego-Ornan E, Fratzl P. Mechanical function of a complex three-dimensional suture joining the bony elements in the shell of the red-eared slider turtle. *MRS Proceedings.* 2009;1187:1105–1187. doi:10.1557/PROC-1187-KK01-05

11. Huang W, Restrepo D, Jung JY, et al. Multiscale toughening mechanisms in biological materials and bioinspired designs. *Advanced Materials.* 2019;31(43):1901561. doi: https://doi.org/10.1002/adma.201901561

12. Peng X, Zhang B, Wang Z, et al. Bioinspired strategies for excellent mechanical properties of composites. *Journal of Bionic Engineering.* 2022;19(5):1203–1228.

13. Udupi SR, Rodrigues LR. Detecting safety zone drill process parameters for uncoated HSS twist drill in machining GFRP composites by integrating wear rate and wear transition mapping. *Indian Journal of Materials Science.* 2016;2016:1–8.

14. Li D, Li Y, Zhou J, Zhao Z. A novel method to improve temperature uniformity in polymer composites microwave curing process through deep learning with historical data. *Applied Composite Materials.* 2020;27:1–17.

15. Tarafdar A, Liaghat G, Ahmadi H, et al. Quasi-static and low-velocity impact behavior of the bio-inspired hybrid Al/GFRP sandwich tube with hierarchical core: Experimental and numerical investigation. *Composite Structures.* 2021;276:114567. doi: https://doi.org/10.1016/j.compstruct.2021.114567

16. Hou B, Liu P, Wang A, Xie J. Fabrication, microstructure and compressive properties of Ti$_2$AlC/TiAl composite with a bioinspired laminated structure. *Vacuum.* 2022;201:111124. doi: https://doi.org/10.1016/j.vacuum.2022.111124

17. Gao F, Zeng Q, Wang J, Ge J, Shen J, Liu S, Liang J. Experimental and numerical study on the flexural mechanical properties of bioinspired composites with suture structures. *Mechanics of Advanced Materials and Structures.* 2023;31:2680–2688. doi:10.1080/15376494.2022.2162644

18. Li Y, Ortiz C, Boyce MC. Stiffness and strength of suture joints in nature. *Physical Review E.* 2011;84(6):62904. doi: 10.1103/PhysRevE.84.062904

19. Goswami KP, Pakshirajan K, Pugazhenthi G. Process intensification through waste fly ash conversion and application as ceramic membranes: A review. *Science of the Total Environment.* 2022;808:151968. doi:10.1016/j.scitotenv.2021.151968

Chapter 11

Recycling and environmental degradation of polyamides

Mohammad Asif Ali, Maiko Okajima,
and Tatsuo Kaneko

11.1 INTRODUCTION

The progress of human society is influenced by the development of human-made polymers such as nylon as a substitute for synthetically produced silk. The modern polymer with the trade name nylon 6,6 is made by polycondensation of adipic acid and 1,6-hexane diamine with a total of 12 carbon atoms in each repeating unit [1]. Polyamide found its application in the bristles, ropes, fishing nets, biomedical applications, and automobile engines [1, 2]. Intermolecular hydrogen bonding enhances the properties of polyamide in such a way that they are resistant to chemicals, are more thermally stable, have good appearance and toughness, and so on [2–4, 5]. Due to the high thermo-mechanical properties and its high molecular weight distribution ratio, a synthetic polyamide which has high PDI (polydispersive index) is very difficult to degrade either environmentally or by microorganism [5]. However, the increase in global warming is of great concern because of the degradability of polyamides leading to the formation of white pollution and greenhouse gases. Now many researchers have been paying attention to reducing global warming by utilizing the renewable resource-based synthetic polyamides and their degradable behaviour [5–8]. Polyamide which is bio-based has a chance to degrade but some of the synthetic polyamides do not degrade quickly, so it affects the ecology by the accumulation of these plastics in the soil and sea, and therefore has led researchers to seek ways to resolve the problem [4–8]: to find ways to recycle the plastics waste to maintain fossil resources [5]. Bio-based polyamides are materials that are derived from renewable resources. However, in the future, bio-based or biodegradable polyamides may be competitive with petroleum-based polyamides and have the potential to reduce the environmental harm. The poly(esters/amides) derived from an α-amino acid; diols, and different kinds of diacid are promising materials for biomedical applications susceptible to hydrolytic or enzymatic degradation [9]. Throughout the photo-ageing process [8–11], degradation mechanism is complex, involving short chain reaction or simultaneous occurrence of the photochemical process and excitation of (NH–CO) and hydrogen abstraction [11]. Bio-based and biodegradable materials can be produced by natural resources available in the environment and are more preferred

DOI: 10.1201/9781003564355-11

compared to the other alternative materials obtained from the petroleum-based feedstocks [7, 10]. These materials can be degraded to sustain the life cycle equation of the carbon cycle. The degradation of polymer means the breaking down of large molecule polymer into smaller molecules of oligomers and monomers which finally converted into water and CO_2. The polyamide materials are partially feasible for degradability, and break down into small parts after the action of microorganisms such as bacteria or genetically altered bacteria, fungi, insects, and alga. The cheaper disposal of plastic through composting into the natural carbon cycle implies photo-oxidation which is crucial to the understanding of the natural aging of aliphatic polyamide. Previously, many researchers have reported the degradability behaviour of a polymer based on the molecular design. The polyamides can be degraded by following thermal, mechanical, chemical hydrolysis, photodegradation, oxidation (environmental degradation), and biodegradation.

Polyamides are less susceptible to biodegradation but some of the bacteria, fungi, and spores are more attracted to the functional group (amide linkage). Some of the polyamide is more susceptible to degradation after functionalization with heteroatom.

11.2 RESULTS AND DISCUSSION

11.2.1 Bio-based monomers as resources for sustainable plastics

The production of bio-based monomers or polymers that originate from renewable resources with specific pathways is more straightforward and more attractive to the production of sustainable plastics [9, 12]. The bio-derived monomer from renewable resources balances the CO_2 level and advantages of a low carbon footprint [10]. The renewable feedstock with recycling capabilities, high resources, and energy-effectiveness of solvent-free melt polycondensation (nylon salt methods) can be used [12, 13]. Some of the bio-derived diacid and diamine are reported that can be used further for polyamide syntheses, as shown in Figure 11.1. Terpenes which are readily available on a large scale have a high potential biomonomer and can be substituted with petrochemicals. Limonene is a kind of terpene which is an inexpensive natural compound derived from renewable resources that form the diamine R or S forms which can be used as a monomer for the bio-based polyamide [15].

11.2.2 Utilization of biomass as a sustainable carbon footprint

The raw materials that are derived from bio-based materials are burned after use to maintain the carbon footprint [15]. Some of the bio-based and biodegradable plastics such as PLA (polylactic acid) are eco-friendly and sustainable [12]. The polyamides (nylon 11) are derived from castor oil under the

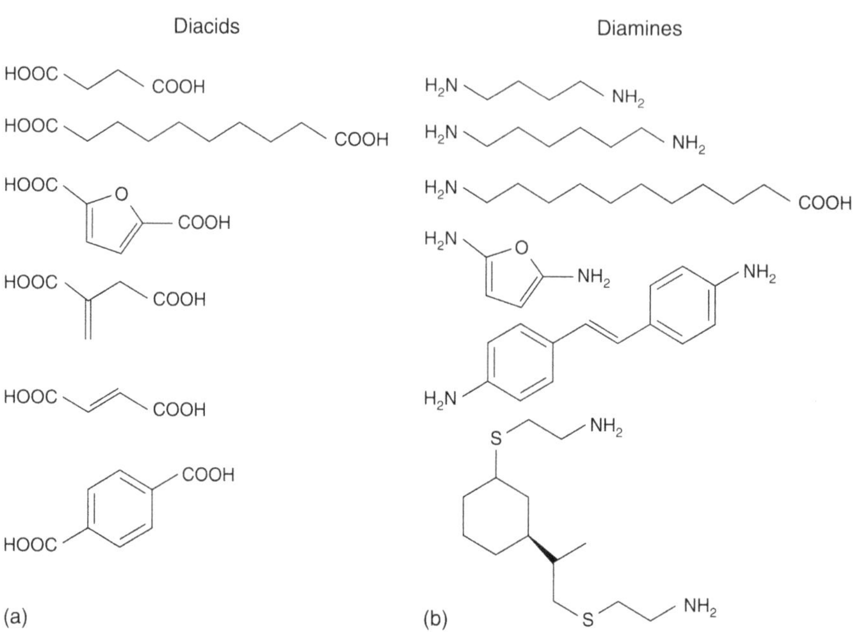

Figure 11.1 Selected renewable monomers supplied by biotechnology.

trade name Rilsan® produced by Arkema but are non-degradable. Some of the polyamides like poly(amino acid), poly(ε- L-lysine), and poly-γ-glutamic are derived from a readily degradable amino acid [1, 2]. The bacteria and spores can be inhaled or absorbed after coming in contact with poly(amino acid). Some of the studies in this area show how these amino acids affect the environment and influence the climatic dependence of biodegradation concerning nanoparticle emissions, which can fix the carbon stocks.

11.2.3 Differences between bio-based and biodegradable polymers

Synthesized bio-based and biodegradable often create confusion in people as both terms imply eco-friendly, but they are not identical. The synthesized bio-based polymers can be obtained either from a natural monomer or by following a synthetic method [10]. The bio-based polymer which shows high performance can be degraded either by a microorganism, fungal infection in soil, or are photodegradable in water, which changes the physical and chemical properties after degradation. Bio-based plastic can be both biodegradable, like poly(lactic acid) (PLA), and non-biodegradable (e.g., biopolyhethylene); however, not all biodegradable polymers are bio-based (e.g., polycaprolactone). Also, the synthetic polyamide nylon 6 is not bio-based, but it is biodegraded by the fungus *Trametes versicolor* NCIM 1086.

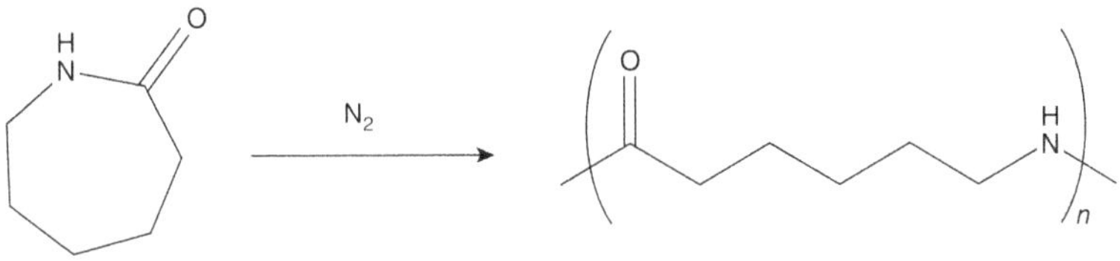

Figure 11.2 Ring-opening polymerization of caprolactam.

11.2.4 Fungal degradation of polyamide 6

Polyamide 6 is a widely used commercial nylon it is synthesized after ring-opening polymerization of ε-caprolactam (Figure 11.2) and is less susceptible to fungal degradation [16]. Polyamide 6 are metabolized by a microorganism such as *Achromobacter, Corynebacterium, Pseudomonas*, and the fungal genera such as *Absidia, Aspergillus, Byssochlamis, Penicillium, Rhodotorula*, and *Trichosporon*. A large number of fungi have been screened only white-rot Basidiomycetes were able to degrade nylon 6.

11.2.5 Enzymatic degradation of polyamides

Some of the PA can degrade after grafting. Grafting of phenolic to polyamides can be done through a two-step enzymatic process. Enzymatic hydrolysis of synthetic fibres by using the hydrolytic enzyme which makes the hydrophobic surface and enzyme was used that secreted from bacteria or fungi. The *Nocardia farcinica* was developed from *Escherichia coli* which secretes the polyamidase hydrolytic enzyme and is applied with laccase. The reaction involves the first hydrolysis of PA with a recombinant polyamide from *N. farcinica* using laccase-catalysed grafting with ferulic acid. For the functionalization of polyamide fibres, ferulic acid and butylamine are required for grafting on PA. This could lead to mild and environment-friendly strategies for the functionalization of PA and can separate the amine functionalities while using both compounds as model substrates. Laccase-mediated reactions between primary amines and phenols have been proposed to proceed via Michael addition or free-radical mechanism. Both led to similar products: in the second step, laccase-catalysed grafting of ferulic acid onto PA was found as suitable for the removal of one functionality, as shown in Figure 11.3. Nylon 6,6 was also treated with different kinds of protease enzymatic with a concentration of the solution (3%, 6%, and 9%). Nylon 610 oligomer (PrePA) was prepared from salt with sebacic acid polyesteramide prepolymer (PrePEAs) having amide content from 10 mol% to 60 mol%. The synthesized PrePEAs through melt polycondensation from adipic acid 1,4-butanediol, and the PrePA with the catalysis of stannous

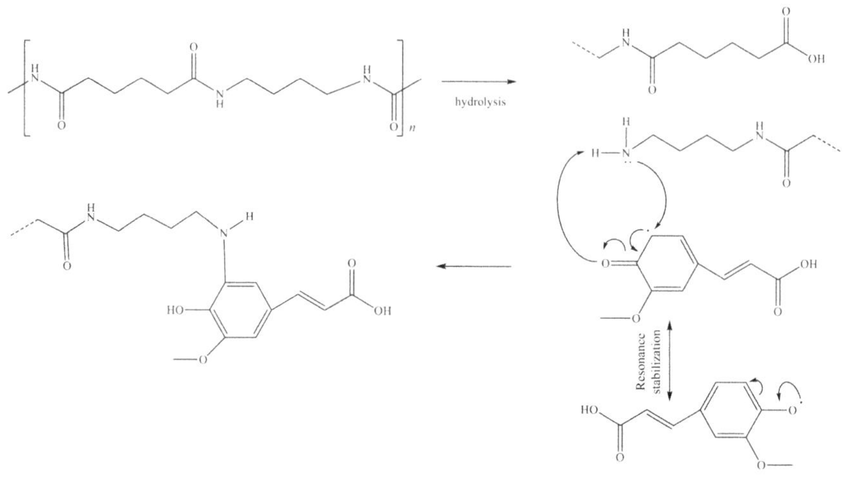

Figure 11.3 Proposed mechanism for two-step enzymatic grafting of polyamide with phenol derivatives.

chloride were susceptible towards the enzymatic degradation (Figure 11.4). The enzymatic degradation of the ExtPEAs from ExtPEA-10 to ExtPEA-60 was conducted at 37°C using proteases. The ring-opening copolymerization of 2-pyrrolidone with ε-caprolactone synthesized hetero-copolymer of copolyesteramides poly(2-pyrrolidone-*co*-ε-caprolactone) which degraded through enzymatic hydrolysis. The biodegradation of PA 4 through microorganisms showed rapid degradation in the amide-rich region. On the contrary, enzymatic hydrolysis using a lipase resulted in a different tendency that is ester-rich copolymers hydrolysed rapidly.

11.2.6 Photo-degradation or photo-stabilization of polyamide

The behaviour of the aliphatic polyamide after exposure to UV radiation may cause the degradation of polyamides. The higher wavelength of the UV light is unable to degrade the polyamide, but the lower wavelength of the UV with

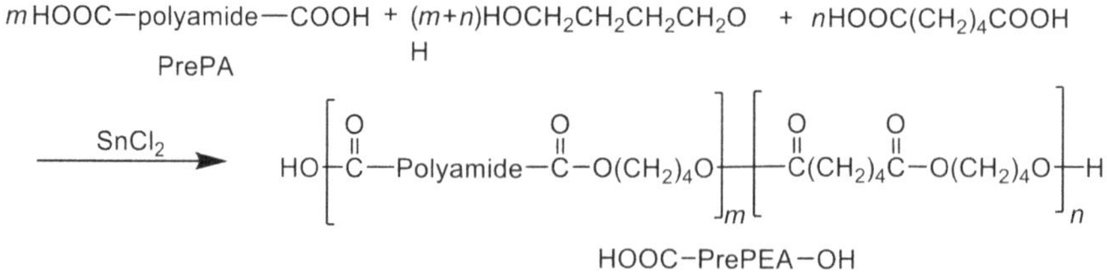

$$m\,HOOC-\text{polyamide}-COOH + (m+n)HOCH_2CH_2CH_2CH_2O\overset{H}{} + n\,HOOC(CH_2)_4COOH$$

PrePA

$$\xrightarrow{SnCl_2} HO\left[\overset{O}{\underset{\|}{C}}-\text{Polyamide}-\overset{O}{\underset{\|}{C}}-O(CH_2)_4O\right]_m\left[\overset{O}{\underset{\|}{C}}(CH_2)_4\overset{O}{\underset{\|}{C}}-O(CH_2)_4O\right]_n H$$

HOOC−PrePEA−OH

Figure 11.4 Synthesis of pre-PEA via polycondensation of pre-PA with butylene glycol and adipic acid.

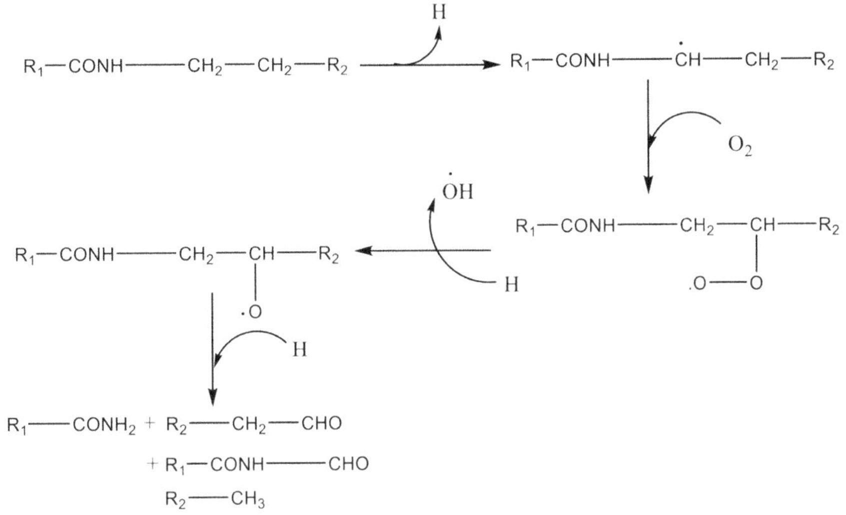

R1—CONH——CH2—CH2—R2
H
R1—CONH——CH—CH2—R2
O2
R1—CONH——CH2—CH—R2
.O—O
OH
H
R1—CONH——CH2—CH—R2
·O
H
R1——CONH2 + R2——CH2——CHO
+ R1—CONH——CHO
R2——CH3

Figure 11.5 The proposed mechanism of polyamide degradation by fungus-derived peroxidase [37].

the photo-oxidation at a wavelength shorter than 290 nm and auto-oxidation favour degradation of the low molecular weight of the polymers (Figure 11.5) [12]. Three-way processes can do the overall photo-oxidation reaction simultaneously occurring in polyamide-6. as summarized in Figure 11.6. Besides, the hydrogen abstraction and subsequent hydro-peroxide formation, two other significant processes appear to be operating in polyamide-6, that is, chain cleavage reactions through Norrish I and Norrish II (Figure 11.6).

CO(CH2)-CH-NH-CO-CH2-CH2-CH2-CH2-CH2-NH-CO
H
hv Norrish I
hv H abstraction
hv Norrish II
CO-NH-(CH)4-CH3
CO-NH-(CH2)4-COOH
CO-NH-(CH2)4-CHO
CO-(CH2)5-NH2
CO-(CH2)5-NH-CHO
CO-NH-(CH2)3-CH=CH2
CO-(CH2)4-CH-NH-
O-OH
CO-(CH2)4-CO-NH-CO
CO-(CH2)4-COOH
CO-(CH2)-CHO
NH-(CH2)5-CO-NH2
CO-(CH2)2-NH-CO-CH
CO-NH-
(CH2)2CH=CH2
CO-NH-(CH2)3-CH3

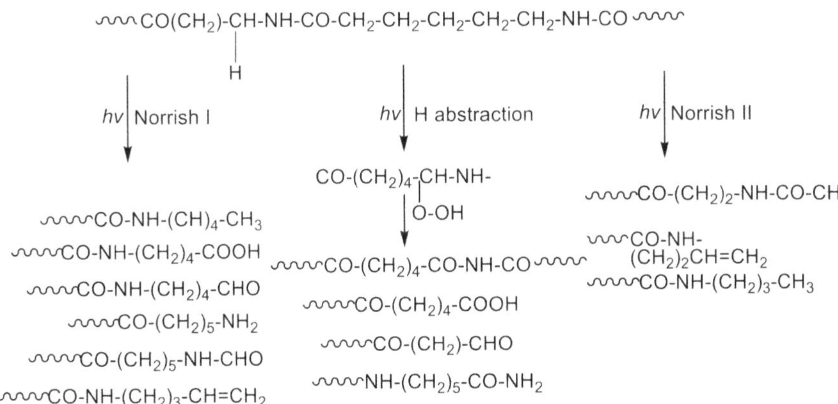

Figure 11.6 Overall photooxidation processes in polyamide 6.

11.2.7 Degradation of poly(amino acid)

The natural bio-based polyamide from the amino acid has been utilized in biomaterials. Synthetic poly(α-amino acid) is also expected to be biodegradable and biocompatible [16, 17]. However, the poly(amino acid) has lower thermomechanical properties but after conjugation with non-amide bonds such as ester, imino carbonates, and carbonates, they are susceptible to degradation [18]. The natural bio-based poly(amino acid) which are composed of one type of amino acid forms poly(ε-L-lysine). However, poly-γ-glutamic acid has α-amide linkages with other types of linkages that involve the β- and γ-carboxylic groups and an ε-amino group. The poly-γ-glutamic is an anionic water-soluble, biodegradable homo-polyamide; the microorganism can degrade it into water and carbon dioxide [15, 19–21]. Some of the bacteria that produce the poly-γ-glutamic acid degraded the membrane protein complex γ-PGA-degrading enzymes by using *Bacillus subtilis*, *B. anthracis*, *Flavobacterium polyglutamicum*, *Myrothecium* sp., and bacteriophages. They are essential to facilitate or antagonize physiological functions of γ-PGA. These bacteria are potential sources of the novel γ-PGA-degrading enzymes [22, 23]. Cyanophycin is a comb-like polypeptide naturally occurring polypeptide from cyanobacteria that can be degraded by extracellular CGPase of the Gram-negative bacterium *Pseudomonas anguilliseptica* BI (CphEPa). The degradation of cyanophycin by *Sedimentibacter hongkongensis* strain KI and *Citrobacter amalonaticus* strain G isolated from an anaerobic bacterial consortium [24, 25]. Poly(L-lysine) is known for antibacterial, antiviral, and antitumour activities, and is degraded by *Sphingobacterium multivorum* [26–29]. Some of the synthetic amino acid polymers are encountered as sophisticated tools for drug delivery, wound-healing, bio-mineralization, and adhesives [27, 28]. There are many synthetic poly(amino acids) like poly(L-glutamic acid) which are prepared through the NCA (*N*-carboxyanhydride) and are structurally different from the γ-PGA, poly(L-glutamic acid) has been found highly liable for degradation by lysosomal enzymes [28, 29].

11.2.8 Degradation of polyamide, including heteroatoms

The molecular design is significant for the deterioration of the polymers, after varying chemical structure which interferes with crystallinity, melting temperature, and the solubility of the polyamide. Functionalities or complexes of this polyamide favour influential property for degradation. A polymer with the ester bond and another heteroatom-containing backbone often undergo enzymatic degradation [29]. However, chemical modification such as the introduction of an alkyl chain provides flexibility and hydrophobicity that are susceptible to degradation. Aliphatic polyamides such as sebacic acid and tetradecanedioic acid containing one or two performed amide linkages are sensitive to degradation [30, 31]. Improved chain flexibility and hydrophobicity were given to

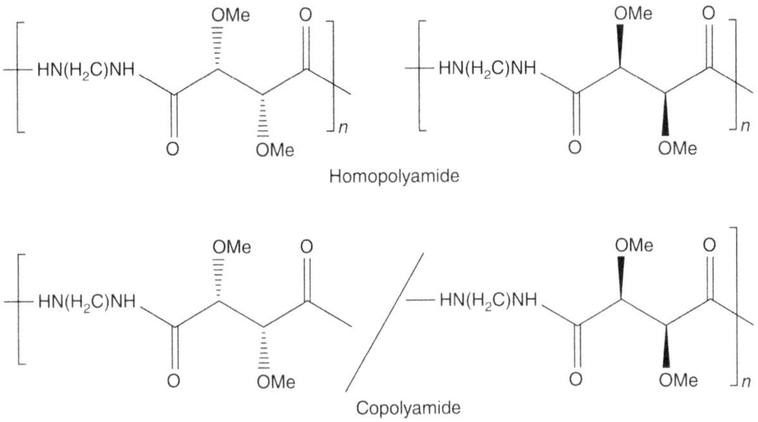

Homopolyamide

Copolyamide

Figure 11.7 Structures of homo-and copolyimide-based D- and L-tartaric acids.

the polyamide using triethyleneoxy-based amide. Stereocopolyamides were synthesized after polycondensation of a mixture of 2,3-di-O-methyl-D- and L-tartaric acids with hexamethylene diamine. However, the varying amount of D/L mixture makes them susceptible to degradation in water under pH 7.4 at 37°C (Figure 11.7). Synthetic poly(ester amide) has been synthesized from 1,6 hexanediol, glycine, and diacid with a various number of methylene groups to enhance the crystalline degradation rate, which is controlled by the amount of the modifier like the phenylalanine:glycine ratio (Figure 11.8) [32]. It can also degrade in the presence of the α-chymotrypsin. The synthetic poly (ester amides) having L-alanine unit are susceptible to the degradation with enzymes (pronase, trypsin, chymotrypsin, and papain) [33].

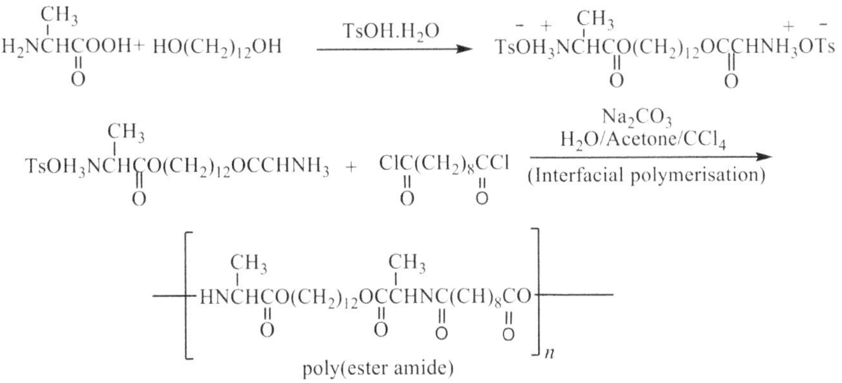

poly(ester amide)

Figure 11.8 Syntheses of poly(ester amide) containing L-alanine unit.

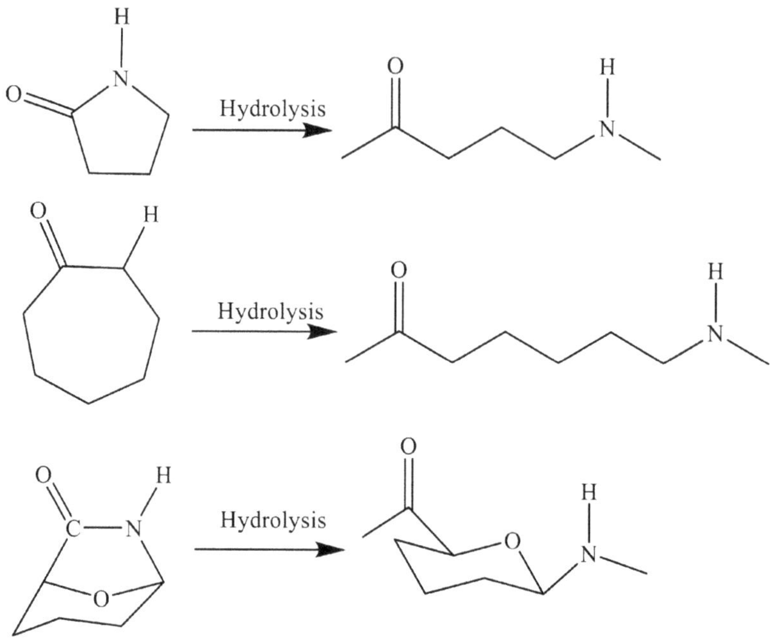

Figure 11.9 Ring-opening conversion of the polymer backbone.

11.2.9 Environmentally corrosive behaviour of polyamides

Synthetic polyamides (nylons) are tricky to decompose in natural conditions through the biological or hydrolytic process due to the high symmetry in the molecular structure and strong interaction of the hydrogen bonding. After chemical modification such as poly(2-pyrrolidone), poly(ε-caprolactam), an itaconic acid-based polypyrrolidone unit is susceptible to the ring-opening. The polyamide film of the poly(2-pyrrolidone) burial test in the soil (pH 7.5) for four months shows a decrease in the molecular weight (Figure 11.9) [33]. The ring-opening hydrolysis of itaconic acid-based polyamide resins in the presence of an alkaline pH >10 shows water-solubility behaviour (Figure 11.10) [34].

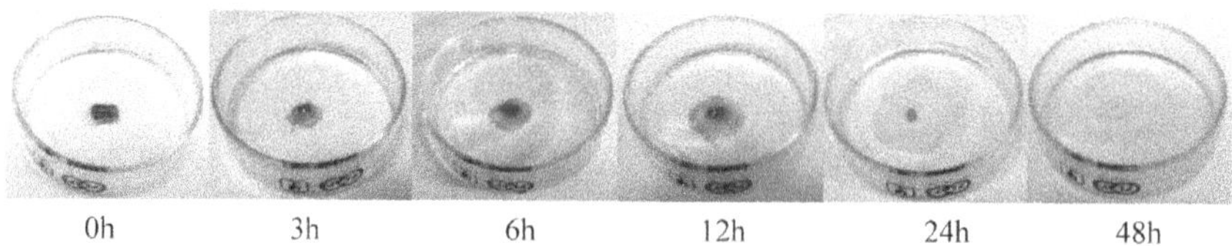

Figure 11.10 Time course of water-solubilization behaviour of resin derived from itaconic acid with 1,5-diaminopentane by a high-pressure mercury lamp.

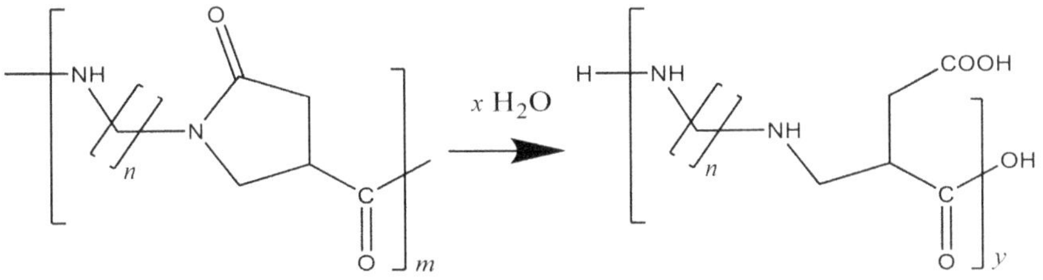

Figure *11.11* Itaconic acid-derived polyamides showing ring-opening conversion from pyrrolidone ring to acid in the polymer backbone.

Photoinduced solubilization of the polyamides in water upon UV irradiation with a high-pressure mercury lamp at a wavelength of 250–450 nm was observed (Figure 11.11). If the polyamides are inside the sea or river, it becomes water-soluble due to a photoreaction under sunlight at 280–400 nm wavelengths; the plastic waste may be corroded. However, it will take a longer time to perform a corrosion test using only direct sunlight. The degradability was also confirmed through the soil-degrading process via weight loss, the resins with sizes of 2–4 cm held by polyethylene nets were buried. After one year, the polymers were recovered, and the shapes and weights of the samples were confirmed, as shown in Figure 11.12. Most of the polyamide samples shrank indicating the corrosion behaviour in the soil. In particular, polyamides derived from itaconic acid with 1,3-diaminopropane, 1,4-diaminobutane and 1,6-diaminohexane disappeared due to in-soil corrosion. Similarly, the polyamides composed of IA with 1,5-diamino pentane (cadaverine) and 1,2-ethylenediamineshowed weight loss values of 96 wt.%, and 98 wt.%, respectively [34]. However, poly(lactic acid) co-buried with the polyamides showed only 16 wt.% weight loss. The aromatic polyamides which are derived from itaconic acid and *m*-xylenediamine, 4,4′-diaminodiphenylether,

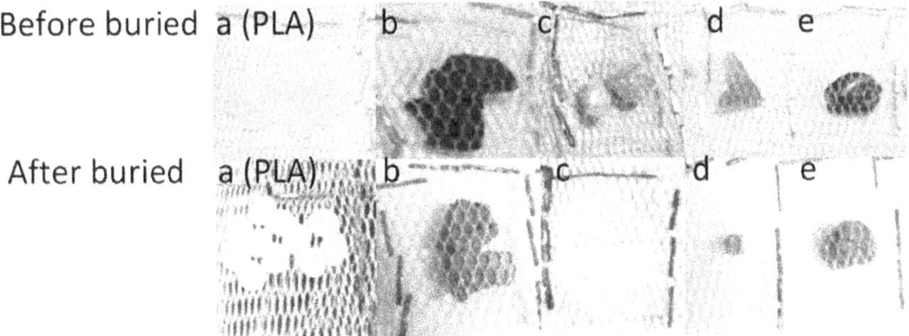

Figure *11.12* Soil corrosion behaviour of itaconic-based polyamides under the soil.

and *p*-phenylenediamine are less susceptible to degradation as compared with aliphatic-based polyamide but are more degraded as compared with poly(lactic acid) [35]. The weight loss of the polymers due to soil corrosion can result from the cumulative effects of physical, chemical, and biological processes.

Polyamide 4 is a linear polymer derived from γ-aminobutyric acid (GABA), which was degraded by the bacterial strain *Pseudomonas* sp. [36].

11.2.10 Limitations and future scope of polyamide

The non-degradable thermoplastics especially Nylon™ are used as packaging materials, fibres, and film forms; some of the polyamides, especially nylon 4, nylon 6, and nylon 6,6, degraded in vivo through some microbes. The chapter has covered significant concerns about the degradation of the polyamide by microorganisms, soil degradation, photo-degradation, photo-solubilization, and degradation via some chemical processes. In general, degradation of polyamide is very difficult—only polyamide 4 can be degraded in general natural environments. Synthetic or bio-derived polyamide such as polyamide-4, polyamide-6, and polyamide-66 follows degradation either via environmental or biological pathways, which manage the different methods of waste disposal. It is approaching to reduce global warming and maintain the carbon footprint, which is much more important for potential application in the fields related to the environment. Itaconic acid-based polyamides are considered to be next-generation bio-based and environmentally degradable heterocyclic polyamides and contribute significantly to establishing a sustainable society. It can be degraded under soil or photo- solubilization behaviour under UV light. Some of the synthetic polyamides derived from renewable resources could be positively impacted by petroleum-based polyamide because of their contribution to the prevention of the global environment, limiting the use of polyamide under water because of amide linkage which absorbs a lot of moisture which slightly influences the degradation ability due to oxidation. For the complete degradation of polyamide or to address polyamide dumping issues, exploring enzymes such as lipase and pepsin for efficient degradation and recovery of monomers, oligomers, or other fragments is a promising approach. Monomers PA-6 and PA-66 can be biodegraded, so there are also some studies that synthesize environmentally degradable polyamides through copolymerization or molecular design, which also provides us with new research ideas. Using a natural filler such as wood fibres in a bio-based polyamide may influence the crystallinity which influence the chemical or physical degradability of polyamides. However, most of the researchers focus on the utilization of bio-based monomers with tailored polymer properties, which is the most promising approach towards bio-based polyamide. Polyamide is more technically difficult to recycle or degrade than the polyester, so finding another alternative way for promising solution in current scenario is paramount.

REFERENCES

1. Jiang Y, Loos K. Enzymatic synthesis of biobased polyesters and polyamides. Polymers 2016;8,7:243.
2. Ishihara K, Ohara S, Yamamoto H. Direct polycondensation of carboxylic acids and amines catalyzed by 3,4,5-trifluorophenylboronic acid. Macromolecules 2000;33:3511.
3. Wang X, Yang W, Li F, Xue Y, Liu R, Hao Y In situ microwave-assisted synthesis of porous n-TiO$_2$/g-C$_3$N$_4$ heterojunctions with enhanced visible-light photocatalytic properties. Industrial and Engineering Chemistry Research 2013;52:17140.
4. Kuo PC, Sahu D, Yu HH. Properties and biodegradability of chitosan/nylon 11 blending films. Polymer Degradation and Stability 2006;91:3097.
5. Nam KT. Solvent degradation of nylon-6 and its effect on fiber morphology of electrospun mats. Polymer Degradation and Stability 2011;1984–1988.
6. Kaneko T, Thi TH, Shi DJ, Akashi M. Environmentally degradable, high-performance thermoplastics from phenolic phytomonomers. Nature Materials 2006;5:966.
7. Friedrich J, Zalar P, Mohorčič M, Klun U, Kržan A. Ability of fungi to degrade synthetic polymer nylon-6. Chemosphere 2007;67:2089.
8. Wendy A, Allan A, Brian T A review of biodegradable polymers: Uses, current developments in the synthesis and characterization of biodegradable polyesters, blends of biodegradable polymers and recent advances in biodegradation studies. Polymer International 1999;47:89.
9. Vroman I, Tighzert L. Biodegradables polymers. Materials 2009;2:307.
10. Iwata T. Biodegradable and bio-based polymers: Future prospects of eco-friendly plastics. Angewandte Chemie 2015;54:3210–5.
11. Klun U, Friedrich J, Kržan A. Polyamide-6 fibre degradation by a lignolytic fungus. Polymer Degradation and Stability 2003;79:99.
12. (a) Firdaus M, Meier MAR. Renewable polyamides and polyurethanes derived from limonene. Green Chemistry 2013;15:370. (b) Deguchi T, Kitaoka Y, Kakezawa M, Nishida T. Purification and characterization of a nylon-degrading enzyme. Applied and Environmental Microbiology 1998;64:1366.ddd
13. Mülhaupt R. Green polymer chemistry and bio-based plastics: Dreams and reality. Macromolecular Chemistry and Physics 2012;214:159–4.
14. Kumar S, Tadahisa I. Sustainability of biobased and biodegradable plastics. CLEAN: Soil, Air, Water 2008;36:433.
15. Ogawa Y, Yamaguchi F, Yuasa K, Tahara Y. Efficient production of γ-polyglutamic acid by *Bacillus subtilis* (natto) in jar fermenters. Bioscience, Biotechnology, and Biochemistry 1997;61:1684.
16. Hassan MK, Mauritz KA, Storey RF, Wiggins JS. Biodegradable aliphatic thermoplastic polyurethane based on poly(ε-caprolactone) and l-lysine diisocyanate. Journal of Polymers Science Part A: Polymer Chemistry 2006;44:2990.
17. Okada M. Chemical syntheses of biodegradable polymers. Progress in Polymer Science 2002;27:87.
18. Fan Y, Chen G, Tanaka J, Tateishi T. Preparation of a biphasic scaffold for osteochondral tissue engineering. Materials Science & Engineering C 2004;24:791.
19. Kimuraand K, Fujimoto Z. Amino-acid homopolymers occurring in nature. Microbilogy Monographs 2010. https://doi.org/10.1007/978-3-642-12453-2_6
20. Kaneko T, Higashi M, Matsusaki M, Akagi T, Akashi M. Self-assembled soft nanofibrils of amphipathic polypeptides and their morphological transformation. Chemistry of Materials 2005;17:2484.
21. Berekaa MM, El-Aassar SA, El-Sayed SM, EL-Borai AM. Production of poly-γ-glutamate (PGA) biopolymer by batch and semicontinuous cultures of immobilized *Bacillus licheniformis* strain-R. Brazilian Journal of Microbiology 2009;40:715.

22. Fan K, Gonzales D, Sevoian M. Hydrolytic and enzymatic degradation of poly(γ-glutamic acid) hydrogels and their application in slow-release systems for proteins. Journal of Environmental Polymer Degradation 1996;4:253.
23. Keitaroum K, Zui F. Enzymatic degradation of poly-gamma-glutamic acid. Amino-Acid Homopolymers Occurring in Nature. 2010. https://doi.org/10.1007/978-3-642-12453-2_6
24. Obst M, Krug A, Luftmann H, Steinbüchel A. Degradation of cyanophycin by *Sedimentibacter hongkongensis* strain KI and *Citrobacter amalonaticus* strain G isolated from an anaerobic bacterial consortium. Applied and Environmental Microbiology 2005;71:3642–3652.
25. Yukuta T, Akira I, Masatoshi K. Development of biodegradable plastics containing polycaprolactone and/or starch. Polymeric Materials: Science and Engineering 1990;63:742–749.
26. (a) Ren K, Ji J, Shen J. Construction and enzymatic degradation of multilayered poly-l-lysine/DNA films. Biomaterials 2006;27:1152–1159. (b) Kobayashi S, Mullen K. Polyamide Syntheses. In: Ali MA (Ed.), Encyclopedia of Polymeric Nanomaterials (pp. 1750–1762). Berlin: Springer, 2015.
27. Pattabiraman VR, Bode JW. Rethinking amide bond synthesis. Nature 2011;480:471.
28. Cheng J, Deming TJ. Controlled polymerization of β-lactams using metal-amido complexes: Synthesis of block copoly(β-peptides). Journal of the American Chemical Society 2001;123:9457–8. https://doi.org/10.1021/ja0110022
29. (a) Saad B, Suter UW. Biodegradable polymeric materials. Encyclopedia of Materials: Science and Technology 2008;551. (b) Yamaoki YFA, Wojcik WBLK, Minoru KTA, Dzwolak D. Poly(L-glutamic acid) and poly(D-glutamic acid): A high-pressure rescue from a kinetic trap. Physical Chemistry Chemical Physics 2012;116:5172
30. Siracusa V, Rocculi P, Romani S, Rosa MD Biodegradable polymers for food packaging: A review. Trends in Food Science & Technology 2008;19:634–643.
31. Pitt CG Poly(ϵ-caprolactone) and its copolymers. Drugs and the Pharmaceutical Sciences 1990;45:71.
32. Kazuhiko H, Tsuyoshi H, Masahiko O. Degradation of several polyamides in soils. Journal of Applied Polymer Science 2003;54:1579–3.
33. (a) Nowak AP, Breedveld V, Pakstis L, Ozbas B, Pine DJ, Pochan D, Deming TJ. Rapidly recovering hydrogel scaffolds from self-assembling diblock copolypeptide amphiphiles. Nature 2002;417:424. (b) Lee CU, Smart TP, Guo L, Epps T, Zhang D. Synthesis and characterization of amphiphillic cyclic diblock polypeptides from *N*-heterocyclic carbene-mediated zwitterionic polymerization of *N*-substituted *N*-carboxyanhydride. Macromolecules 2011; 44:9574.
34. Ali MA, Tateyama S, Oka Y, Okajima M, Kaneko D, Kaneko T. High-performance biopolyamides derived from itaconic acid and their environmental corrosion. Macromolecules 2013;46:3719–5.(b) Ali MA, Tandon N, Kaneko T. Simultaneous hardening/ductilizing effects of cryogenic nanohybridization of biopolyamides with montmorillonites. ACS Omega 2017;2:9103.
35. Ali MA, Tateyama S, Kaneko T. Synthesis of rigid-rod but degradable biopolyamides from itaconic acid with aromatic diamines. Polymer Degradation and Stability 2014;109:367.
36. Yamano N, Kawasaki N, Ida S, Nakayama Y, Nakayama A. Biodegradation of polyamide 4 in vivo. Polymer Degradation and Stability 2017;137:281.
37. Nobuhiko N, Tetsuya D, Yukie S-A, Toshiaki N-K, Tadaatsu N. Gene Structures and Catalytic Mechanisms of Microbial Enzymes Able to Biodegrade the Synthetic Solid Polymers Nylon and Polyester Polyurethaoe. Biotechnology and Genetic Engineering Reviews 2001;18:1, 125–47, DOI: 10.1080/02648725.2001.10648011.

Index

3D printing, 43

A

B

C

D

E

F

G

H

I

L

M